T0344849

Climate Change Adaptation in Small Island Developing States

Climate Change Adaptation in Small Island Developing States

Martin J. Bush
Toronto, Canada

WILEY Blackwell

This edition first published 2018
© 2018 John Wiley & Sons Ltd

The right of Martin J. Bush to be identified as the author of this work has been asserted in accordance with law.

Registered Office(s)
John Wiley & Sons, Inc., 111 River Street, Hoboken, NJ 07030, USA
John Wiley & Sons Ltd, The Atrium, Southern Gate, Chichester, West Sussex, PO19 8SQ, UK

Editorial Office
9600 Garsington Road, Oxford, OX4 2DQ, UK

For details of our global editorial offices, customer services, and more information about Wiley products visit us at www.wiley.com.

Wiley also publishes its books in a variety of electronic formats and by print-on-demand. Some content that appears in standard print versions of this book may not be available in other formats.

Library of Congress Cataloging-in-Publication data applied for

ISBN: 9781119132844

Cover Design: Wiley
Cover Image: © AnnWorthy/iStockphoto

Set in 10/12pt Warnock by SPi Global, Pondicherry, India
Printed and bound in Malaysia by Vivar Printing Sdn Bhd

10 9 8 7 6 5 4 3 2 1

This book is dedicated to all the young folk who have inherited a wounded and ailing planet. A huge responsibility has been placed upon your shoulders. My hope is that this book, and many others like it, will help you to find the courage and the strength to fight for the survival of planet Earth.

And to Anna Delia Jean, Michael, and Corry who have been very supportive and patient.

Contents

Preface

This book is about climate change and how small island states need to adapt to a climate that is rapidly evolving in ways that presents a multitude of existential threats.

As a group, the Small Island Developing States, the SIDS, produce relatively small amounts of greenhouse gases. There are one or two exceptions to this rule, but for nearly all the islands the quantities of carbon dioxide generated are miniscule. Even so, most of the small island states have committed scarce financial and human resources to reducing their emissions even further as part of their commitments under the Paris Agreement. This is a mistake.

The urgent priority for small island states is to find ways to adapt to the changing climate. Reducing their greenhouse gas emissions doesn't help anyone on the islands, and doesn't even do anything much for the planet.

For many small island communities, the changing climate threatens not just their way of life but their very existence. As sea level rises and storms intensify, many of the SIDS will become uninhabitable as groundwater resources become contaminated by seawater intrusion, and overwashed arable land is rendered infertile by the salinized soil. At the same time, fishing is becoming less productive for many coastal communities as coral reefs are bleached out, and wamer and more acidic ocean waters play havoc with marine species' environment.

Many adaptation strategies exist for small islands. The majority will survive because most of their land is substantially above sea-level. But they will lose much of their coastal areas to the rising oceans, and many coastal communities will be forced to move to higher ground. States that are multi-island countries that include many islands and atolls will have to deal with entire communities that will elect to move *en masse* to another, more secure, neighbouring island. Managing the social disruption and conflict that will result from these migrations will be a major challenge for many island states.

The management of water resources is crucial. As temperatures rise and periods of drought become longer and more frequent, the abundant rainfall of intense storms must be collected, stored, and managed. Agriculture must adapt to a more variable and unpredictable climate, and become more efficient in the way it uses water. Where coastal tourism is an important component of the economy, ways must be found to provide reliable electrical power and adequate potable water for tourism infrastructure.

Strengthening the resilience of coastal and marine ecosystems is a top priority. Ecosystem-based adaptation is always a no-regret action that should be implemented in collaboration with coastal communities.

The one area where climate change mitigation and adaptation go hand in hand is in the transition away from carbon-based fuels to renewable energy. Transitioning electrical power production to renewables such as photovoltaic electricity and wind power brings many important co-benefits that strengthen the resilience of island communities, and enable them to cope better with the extreme weather that is a dominant feature of the changing climate.

Small island states are in an extremely vulnerable position. Only a few of the 51 small island states are wealthy; several are still developing countries. Most of them lack the human and financial resources necessary to effectively tackle the difficulties they face. Coordination among key ministries is poor, and planning is generally weak, unfocused, and inadequate.

As this book went to press, three major hurricanes swept though the Caribbean destroying almost everything in their path. Harvey, Irma and Maria caused enormous damage to many small Caribbean islands. It is a reminder once again that for most of the small island states, focusing on reducing carbon emissions is not the priority. Ways have to be found to adapt to an increasingly unpredictable and often violent climate.

Although the outlook for small island developing states is not good, there is much that can be done if governments are better organized, and ministries cooperate and coordinate their plans. Many islands simply lack effective leadership. This book is intended to help.

Abbreviations and Symbols

AIMS	Africa, Indian Ocean, Mediterranean and South China Sea: a group of the SIDS
bbl	Barrel: 42 US gallons (about 159 litres)
CA	Conservation agriculture
CARICOM	Caribbean Community
CC	Climate change
CDC	Centers for Disease Control (USA)
CO_2	Carbon dioxide
COP	Conference of the Parties (to the UNFCCC)
CSA	Climate-smart agriculture
DA	Designated authority
DR	Dominican Republic
EBA	Ecosystem-based adaptation
EEZ	Exclusive economic zone
ENSO	El Nino Southern Oscillation
ESM	Earth system model
EVI	Economic Vulnerability Index
GCF	Green Climate Fund
GCM	General circulation model
GDP	Gross domestic product
GHGs	Greenhouse gases (principally CO_2, methane and nitrous oxide)
GIS	Geographical information system
Gt	Gigatonne (1 billion tonnes)
GtC	Gigatonne of carbon
$GtCO_2$	Gigatonne of carbon dioxide
$GtCO_2e$	Gigatonne of carbon dioxide equivalent (includes other greenhouse gases)
GW	Gigawatt (1 billion watts)
GWh	Gigawatt-hour (1 billion watt-hours)
HDR	Human development report
IAM	Idealized assessment model
IFPRI	International Food Policy Research Institute
INDC	Intended Nationally Determined Contribution
IPCC	Intergovernmental Panel on Climate Change

IPP	Independent power producer
IUCN	International Union for the Conservation of Nature
kWp	Kilowatt peak
LDC	Less developed country
LECZ	Low elevation coastal zone
LED	Light-emitting diode
MENA	Middle East and North Africa
MPA	Marine protected area
MSL	Mean sea-level
$MtCO_2$	Million tonnes of CO_2
MW	Megawatt (1 million watts)
MWh	Megawatt-hour (1 million watt-hours)
MWp	Megawatt peak
NASA	National Aeronautical and Space Agency (USA)
NDA	National designated authority
NHC	National Hurricane Centre (USA)
NOAA	National Oceanic and Atmospheric Administration (USA)
pH	A measure of acidity (pH of 7 is neutral, a lower value is more acidic)
PN3B	Three Bays National Park (in Haiti)
PNG	Papua New Guinea
PPA	Power purchase agreement
ppb	Parts per billion
ppm	Parts per million
PV	Photovoltaic (energy)
RCP	Representative concentration pathway
RF	Radiative forcing
SIDS	Small Island Developing State
SST	Sea surface temperature
t/yr	Tonnes per year
UNDESA	United Nations Department of Economic and Social Affairs
UNFCCC	United Nations Framework Convention on Climate Change
WHO	World Health Organization

1

The Changing Climate

Introduction

This introductory chapter outlines and summarizes the latest information and data about the Earth's changing climate. It relies to a large extent on the fifth Assessment Report of the Intergovernmental Panel on Climate Change – the IPCC, the international scientific agency that reports every four or five years on climate change. But the chapter also integrates much of the most recent information on the impact of climate change, some of which suggests that the IPCC underestimates the threat to human welfare across the globe. The aim of the chapter is to look at the big picture in terms of the global impact of climate change. In subsequent chapters we will look at the impact of climate change on the different sectors of a country's economy, and then specifically how climate change is an increasingly dangerous threat for Small Island Developing States (SIDS), and what measures can be taken to reduce the level of that threat.

The scientific evidence that human activity has influenced the climate system is overwhelming. The climate is changing and in ways that have never before been experienced in human history. The atmosphere and the oceans are warmer, continental areas of snow and ice have diminished, and sea-levels have risen. These are well-established scientific facts. Reliable climate data show that each of the last three decades has been successively warmer at the surface of the Earth than any preceding decade since measurements began over 150 years ago.

The evidence shows that the three decades before 2012 were the warmest period over several centuries in the northern hemisphere, and quite possibly the warmest period in more than a thousand years. Data measured by NASA and NOAA confirmed that 2014, and then 2015, were the hottest years on record. Then 2016 broke those records again. The year 2016 was the warmest on record in all the major global surface temperature datasets (NASA, 2015a; WMO, 2017).

The cryosphere is undergoing a huge transition: snow cover, sea ice, lake and river ice, glaciers, ice caps and ice sheets, permafrost and seasonally frozen ground, are all thawing and melting. Glaciers are melting almost everywhere and have contributed to sea-level rise throughout the twentieth century. The rate of ice loss from the Greenland ice sheet has substantially increased over the last 20 years. Melting from the Antarctic ice sheet, mainly from the northern Antarctic peninsula and the Amundsen Sea sector of West Antarctica, has also increased. The extent of Arctic sea ice has decreased in every season, with the most rapid decrease taking place every summer. The trend

Climate Change Adaptation in Small Island Developing States, First Edition. Martin J. Bush.
© 2018 John Wiley & Sons Ltd. Published 2018 by John Wiley & Sons Ltd.

continued in 2017 with the extent of the sea ice at both poles dropping to record levels. Never before in the satellite records has the area of sea ice at the north and south poles simultaneously fallen so dramatically. The summer Arctic sea ice minimum is decreasing by about 10–13% per decade – a figure that translates to around one million km^2 each decade.

Snow cover has decreased in the northern hemisphere since the middle of the last century. In addition, because of the higher surface temperatures and changing snow cover, permafrost temperatures have increased in the northern hemisphere with commensurate reductions in thickness and area.

Figure 1.1 shows the trend in global mean temperatures since 1880 (NASA, 2015b).

More than 90% of the thermal energy accumulated in the climate system over the last couple of decades has been absorbed and stored in the oceans. Only about 1% of this heat is held in the atmosphere.

Tracking ocean temperatures and the associated changes in ocean heat content allows scientists to monitor variations in the Earth's energy imbalance. Ocean waters are getting warmer: the effect is greatest near the surface, and the upper 75 metres have been warming by over 0.1 °C per decade (IPCC, 2014a). But not only warmer: many large geographical areas of ocean water are becoming more saline as evaporation increases due to the higher surface temperatures. In contrast, other ocean areas, where precipitation is the dominant water cycle mechanism, may have become less saline.

These regional and differing trends in ocean salinity provide indirect evidence for widespread changes in evaporation and precipitation over the oceans, and by extension in the global hydrological cycle. These changes have major implications for rainfall patterns and intensities worldwide, and also for global patterns of ocean water circulation. As the lower atmosphere becomes warmer, evaporation rates increase, resulting in an increase in the amount of water vapour circulating throughout the troposphere. A consequence of this phenomenon is an increased frequency of intense rainfall events,

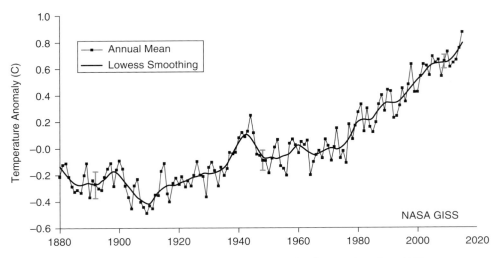

Figure 1.1 Global mean temperature changes based on land and ocean data since 1880.
Source: Courtesy of NASA (2015b), http://data.giss.nasa.gov/gistemp/graphs/.

mainly over land areas. In addition, because of warmer temperatures, more precipitation is falling as rain rather than snow – which has consequences for regional patterns of spring runoff.

As the oceans warm they expand, resulting in both global and regional sea-level rise. The increased heat content of the oceans accounts for as much as 40% of the observed global sea-level rise over the past 60 years.

The slow but steady change in the global water cycle has also had an impact on sea-levels worldwide. Over the last century, global mean sea-level rose by about 0.2 metres. The rate of sea-level rise is also increasing: the rate now is greater than at any time during the last two millennia. NASA satellites have shown that sea-levels are now rising at about 3 mm a year: a total of more than 50 mm between 1993 and 2010 (NASA, 2015c).

Some regions experience greater sea-level rise than others. The tropical western Pacific saw some of the highest rising sea-level rates over the period 1993–2015 – which became a significant factor in the extensive devastation of areas of the Philippines when typhoon *Haiyan* generated a massive storm surge in November 2013 (WMO, 2017).

The absorption of carbon dioxide (CO_2) by ocean seawater, driven by higher atmospheric concentrations of the gas, has resulted in an increase in the acidity of the oceans. The acidity (pH) of ocean surface water has decreased by 0.1, which corresponds to a 26% increase in acidity, a change that many marine species cannot endure. In addition, as a result of the warming trend, oxygen concentrations have decreased in coastal waters and in many ocean regions.

Any changes in the Earth's climate system that affect how much energy enters or leaves the Earth and its atmosphere alters the Earth's energy equilibrium and will cause global mean temperatures to rise or fall. These changes, called radiative forcings (RF), quantify the variations in the amount of energy in the Earth's climate system. Natural climate forcings include changes in the sun's brightness, Milankovitch cycles (small variations in the Earth's orbit and its axis of rotation), and large volcanic eruptions that inject dust and particulates high into the atmosphere and reduce incoming solar radiation.

However, the largest contributor to radiative forcing by far is the concentration of greenhouse gases (GHGs) in the atmosphere. Greenhouse gas emissions caused by human activities have increased markedly since the pre-industrial era, driven largely by economic and population growth. From 2000 to 2010, GHG emissions were the highest in history, and have driven atmospheric concentrations of carbon dioxide, methane, and nitrous oxide to levels unprecedented in at least the last 800,000 years. Concentrations of carbon dioxide, methane, and nitrous oxide have all seen particularly large increases over the period from 1750 to the present day (IPCC, 2014a).

- Carbon dioxide (CO_2) levels have risen from about 280 parts per million (ppm) to 400 ppm in 2015.
- Methane has more than doubled, rising from 700 per billion (ppb) to more than 1800 ppb.
- Nitrous oxide has risen from about 270 ppb to more than 320 ppb.

At the beginning of this century, carbon dioxide concentrations increased at the fastest observed decadal rate of change. For methane, after almost a decade of stable concentrations since the late 1990s, atmospheric measurements have shown renewed

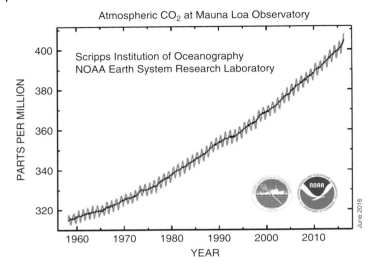

Figure 1.2 Atmospheric CO_2 concentrations measured by the NOAA since before 1960. *Source:* NOAA (2015). Courtesy of NOAA ESRL Global Monitoring Division.

increases since 2007. Nitrous oxide concentrations have also increased steadily over the last three decades.

Since 1970, cumulative CO_2 emissions from fossil fuel combustion, cement production and flaring have tripled, while emissions from forestry and land use changes have increased by about 40%. Figure 1.2 tracks how emissions of CO_2 have been constantly climbing for the last 60 years.

Carbon dioxide is the predominant greenhouse gas, accounting for about three-quarters of total GHG emissions. According to the IPCC, since the beginning of the industrial era about 2000 billion tonnes (Gt) of CO_2 have been released into the Earth's atmosphere (IPCC, 2014a). Of this total, approximately 40% of CO_2 emissions remain in the atmosphere; the remainder is removed from the atmosphere by sinks or stored in natural carbon cycle reservoirs. Ocean absorption and storage in vegetation and soils account in about equal measure for the remainder of the CO_2 emissions: the oceans absorb about 30% of the emitted CO_2, which is what causes the increase in the acidity of ocean seawater.

Figure 1.3 shows how carbon dioxide moves around in the global carbon cycle (LeQueré *et al.*, 2016). The numbers, in gigatonnes of carbon dioxide per year (Gt/yr), are averaged over the decade 2006–2015. The up arrows show emissions from fossil fuels and industry, and from land-use change; the down arrows indicate carbon dioxide that is absorbed by the 'sinks': land and the oceans. The excess carbon dioxide remains in the atmosphere where it is accumulating constantly, as Figure 1.2 confirms.[1]

In the twenty-first century, emissions of GHGs have increased substantially, with larger absolute increases between 2000 and 2010. In spite of the increasing number of climate change mitigation policies implemented worldwide, annual GHG emissions

---

1 All fluxes shown in Figure 1.3 are in units of Gt/yr with uncertainties reported as ±1 standard deviation. So there is a 68% confidence that the real value lies within the given interval (see LeQueré *et al.*, 2016).

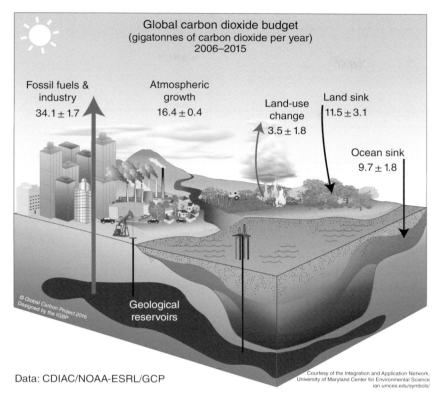

Figure 1.3 Global carbon dioxide budget over the period 2006 to 2015. *Source:* LeQueré *et al.* (2016), www.earth-syst-sci-data.net/8/605/2016/. Used under CC-BY-3.0. (*See insert for color representation of the figure.*)

grew on average by 2.2% annually from 2000 to 2010 compared with 1.3% per year from 1970 to 2000. Recent GHG emissions are the highest in human history; the global economic crisis of 2007/2008 reduced emissions only temporarily (IPCC, 2014a). Total anthropogenic GHG emissions from 2000 to 2010 were the highest ever recorded, reaching 49 GtCO$_2$e in 2010 and then rising to over 52 GtCO$_2$e in 2014 (UNEP, 2016).

Greenhouse gas emissions have levelled off since about 2012, due to the shift to cleaner carbon fuels, huge investments in renewable energy, and the increased efficiency in the way energy is used, but the present level of emissions is still far above the much lower levels needed to keep the Earth's atmospheric temperatures below the Paris target of 2 °C. Emissions need to not just flatline: they need to fall precipitously. There is no sign yet that this is happening.

Although radiative forcing has been increasing at a lower rate over the period 1998 to 2011, this is partly due to cooling effects from volcanic eruptions and the cooling phase of the solar cycle between 2000 and 2009. For the period from 1998 to 2011, most climate model simulations show a surface warming trend that is greater than the observations. This difference is thought to be caused by natural internal climate variability, which sometimes enhances and sometimes counteracts the long-term forced warming trend. Natural internal variability can therefore reduce the reliability of short-term trends in long-term climate change. However, for the longer period between 1951 and 2012, simulated surface warming trends are consistent with observed values.

Recent Impacts of Climate Change

The global climate has been changing for the last several decades, but the changes have become increasingly evident, recorded in more detail and better understood. Mean global surface temperatures are now more than 1°C higher than they were in pre-industrial times, and impacts from climate change have been recorded on natural and human systems on all continents and across all the oceans. The impact of climate-related extremes include the degradation of ecosystems, the disruption of food production and water supply, damage to infrastructure and human settlements, increased sickness and mortality, and negative effects on mental health and wellbeing. The information below summarizes observed climate-change related events over the last few years for seven regions of the world (IPCC, 2014b; WMO, 2017).

In Africa, recent observations attributed to climate change include:

- Extreme heatwaves and drought in southern Africa from late 2015 to early 2016.
- The retreat of tropical highland glaciers in East Africa.
- Lake surface warming and water column stratification increases in the Great Lakes and Lake Kariba.
- Increased soil moisture drought in the Sahel since 1970, although partially wetter conditions since 1990.
- Tree density decreases in western Sahel and semi-arid Morocco.
- Range shifts of several southern plants and animals.
- Decline in coral reefs in tropical African waters.

In Europe:

- Increasingly frequent heatwaves (temperatures in Spain reached 45.4°C in September 2016).
- The retreat of Alpine, Scandinavian and Icelandic glaciers.
- Increase in rock slope failures in the western Alps.
- Earlier greening, leaf emergence, and fruiting in temperate and boreal trees.
- Increased colonization of alien plant species.
- Earlier arrival of migratory birds since 1970.
- Increasing burnt forest areas during recent decades in Portugal and Greece.
- Northward distributional shifts of zooplankton, fishes, seabirds and benthic invertebrates in NE Atlantic.
- Northward and depth shift in distribution of many fish species across European seas.
- Plankton phenology changes in NE Atlantic.
- Spread of warm water species into the Mediterranean beyond changes due to invasive species and human impacts.
- Impacts on livelihoods of Sami people in northern Europe.
- Stagnation of wheat yields in some countries in recent decades despite improved technology.
- Positive yield impacts for some crops mainly in northern Europe.
- Spread of bluetongue virus in sheep and ticks across parts of Europe.

In Asia:

- Permafrost degradation in Siberia, Central Asia and the Tibetan plateau.
- Shrinking mountain glaciers across most of Asia.
- Increased flow in several rivers due to shrinking glaciers.
- Earlier timing of spring floods in Russian rivers.
- Reduced soil moisture in northern China (1950–2006).
- Surface water degradation in parts of Asia.
- Changes in plant phenology and growth in many parts of Asia.
- Distribution shifts of many plant and animal species upwards in elevation or poleward.
- Advance of shrubs into the Siberian tundra.
- Decline in coral reefs in tropical Asian waters.
- Northward range extension of corals in the East China Sea and western Pacific, and of predatory fish in the Sea of Japan.
- Negative impacts on aggregate wheat yields in south Asia.

In Australasia:

- Significant decline in late-season snow depth at three out of four alpine sites (1957–2002).
- Substantial reduction in ice and glacier ice volume in New Zealand.
- Reduced inflow in river systems in SW Australia since mid-1970s.
- Changes in genetics, growth, distribution and phenology of many species particularly birds, butterflies and plants.
- Expansion of monsoon rainforest at the expense of savannah and grasslands in northern Australia.
- Changed coral disease patterns at the Great Barrier Reef and widespread coral bleaching.
- Advanced timing of wine grape maturation in recent decades.

In North America:

- Shrinkage of glaciers across western and northern areas.
- Decreasing amount of water in spring snowpack in western areas.
- Shift to earlier peak flow in snow-dominated rivers in western areas.
- Phenology changes and species distribution shifts upward in elevation and northward across multiple taxa.
- Increased wildfire frequency in subarctic conifer forests and tundra.
- Northward distributional shifts of northwestern fish species.
- Changes in mussel beds along the west coast.
- Changed migration and survival of salmon in northeast Pacific.
- Increased coastal erosion in Alaska and Canada.
- Impacts on livelihoods of indigenous groups in the Canadian Arctic.

In Central and South America:

- Shrinkage of Andean glaciers.
- Changes in extreme flows in the Amazon River.
- Changing discharge patterns in rivers in the western Andes.
- Increased streamflow in sub-basins of the La Plata River.
- Increase coral bleaching in the western Caribbean.

- More vulnerable livelihood trajectories for indigenous Aymara farmers in Bolivia due to water shortages.
- Increase in agricultural yields and expansion of agricultural areas in southeastern South America.

In the polar regions, changes attributed to climate change are widespread:

- Decreasing Arctic sea ice cover in summer.
- Reduction in ice volume in Arctic glaciers.
- Decreasing snow cover extent across the Arctic.
- Widespread permafrost degradation especially in the Southern Arctic.
- Ice mass loss across coastal Antarctica.
- Increased winter minimum river flow in most of the Arctic.
- Increased lake water temperatures and prolonged ice-free seasons.
- Disappearance of thermokarst lakes due to permafrost degradation in the low Arctic.
- New lakes created in areas of formerly frozen peat.
- Increased shrub cover in tundra in North America and Eurasia.
- Advance of Arctic tree line in latitude and altitude.
- Changed breeding area and population size of subarctic birds due to snowbed reduction and/or tundra shrub encroachment.
- Loss of snowbed ecosystems and tussock tundra.
- Impacts on tundra animals from increased ice layers in snow pack, following rain-on-snow events.
- Increased plant species ranges in the West Antarctic Peninsular and nearby islands over the past 50 years.
- Increased phytoplankton productivity in Signy Island lake waters.
- Increased coastal erosion across Arctic.
- Negative effects on non-migratory Arctic species.
- Decreased reproductive success in Arctic seabirds.
- Reduced thickness of foraminiferal shells in southern oceans due to ocean acidification.
- Reduced krill density in Scotia Sea.

All the observations cited above are those where IPCC scientists are confident that they are primarily due to climate change, and not to other local and regional factors that have impacted the environment and ecosystems.

Reports From the Front Line

In 2015, 189 countries filed communications with the UNFCCC secretariat in advance of the 21st Conference of Parties (COP 21) held in Paris in December of that year. One hundred and thirty-five of those countries included information about their adaptation programmes, the climate change impacts they were already experiencing, and their assessment of their vulnerability to the way the climate is changing (UNFCCC, 2016).

In terms of observed changes, many countries reported on the temperature increases in their territories, ranging from around 0.5 to 1.8 °C since the 1960s. Others referred to the rate of change of temperature per annum or over decades. Some countries referred to observed sea-level rise, ranging from 10–30 cm in the past 100 years, or 1.4–3 mm per year.

Other observed changes highlighted by many countries included: increased frequency of extreme weather, in particular floods and drought; changes, mostly negative, in rainfall patterns; and increased water scarcity. For instance, one country reported that water availability per capita is now three times lower than in 1960, while another country indicated that annual maximum rainfall intensity in one hour increased from 80 mm in 1980 to 107 mm in 2012. One country reported that some of the islands in its territory have disappeared under water, while another referred to the near-disappearance of Lake Chad.

Most of the communications submitted by these countries contain a description of the key climate hazards faced by the countries concerned. The three main sources of concern are flooding, droughts and higher temperatures. Many countries have observed extreme weather such as stronger wind and rain, cyclones, typhoons, hurricanes, sea storm surges, sandstorms and heatwaves. Countries also mentioned slow-onset impacts such as ocean acidification and coral bleaching, saltwater intrusion and changes in ocean circulation patterns, desertification, erosion, landslides, vector-borne disease, as well as the high risk of glacial lake outburst floods.

The most vulnerable sectors most referred to by all the countries were water, agriculture, biodiversity and health. Forestry, energy, tourism, infrastructure and human settlement are also identified as vulnerable by a number of countries, and wildlife was mentioned by at least three.

In terms of the most vulnerable geographical zones: arid or semi-arid lands, coastal areas, river deltas, watersheds, atolls and other low-lying territories, isolated territories and mountain ranges are all identified in the reports, and some countries identified specific regions that were most vulnerable. Two countries stressed that they were at risk of losing significant amounts of economically important land in river deltas due to sea-level rise.

Vulnerable communities were identified as being mostly composed of rural populations, in particular smallholders, women, youth and the elderly. Several countries provided quantitative estimates of vulnerable people or communities, sometimes using specific indicators. One country identified 319 municipalities as highly vulnerable; another categorized 72 of its 75 districts as highly vulnerable and identified specific risks. One country stated that 42 million people might be affected by sea-level rise due to its long coastline.

In addition to climate impacts, many countries referred to the social, economic and political consequences of a changing climate. Many referred to the risk of fluctuating food prices and other related risks such as the declining productivity of coral reef systems, reduced crop yields or fishing catches, as well as to water security challenges due to scarcity or contamination. For instance, one country stated that that the flow of the Nile is projected to decrease by 20–30% in the next 40 years, creating serious water supply concerns. Others are concerned about the loss of pastoral land, and some countries fear that changes in precipitation and the growing season may disrupt their agricultural calendars. Other drew attention to specific threats to infrastructure and property. In this context, a few countries drew attention to concerns for social justice, stressing that high-risk areas are often populated by the poorest and most marginalized segments of society. A few countries are recovering from conflicts and indicated that climate change poses an additional burden on their fragile state. Two countries highlighted that water scarcity has triggered conflicts between nomadic peoples or pastoral communities (UNFCCC, 2016).

The Caribbean drought of 2009–2010

Perhaps linked to a strong El Nino in 2009, the drought impacted all the islands from the Bahamas down to Guyana. St Lucia declared a water emergency after its main reservoir's level dropped more than six metres. Two schools and several courtrooms were forced to close because of dry taps. In Guyana, a grass-roots women's organization staged a protest and fundraiser to purchase a water truck. In Jamaica, where the island's largest dams had been operating at less than 40% capacity, inmates at a maximum security prison protesting the lack of water started a riot that injured 23 people. Incidents of water theft, illegal connections and vandalism shot up. Trinidad and Tobago enforced a strict water conservation law for the first time in 20 years. In Barbados, crews battled more than 1000 bush fires – nearly triple the number of the previous year.

Future Shock

Scientists have been building mathematical models of the Earth's geophysical and socioeconomic systems since computers became capable of handling the mathematics and data storage requirements back in the late 1960s. Perhaps the most famous global model was the World3 model developed by Donella Meadows and her colleagues, which led to the publication of the book *The Limits to Growth* in 1972. Since that time, mathematical models of global and regional geophysical systems have become much more complex and now require massive amounts of computing power.

The IPCC climate models are mathematical representations of the geophysical processes that drive the Earth's climate system. The models range from simple idealized models to comprehensive general circulation models (GCMs), including Earth system models (ESMs) that simulate the carbon cycle. The models are extensively tested against historical observations to confirm their accuracy, and to enable adjustments of their parameters if their accuracy falls short.

Climate models perform well in reproducing observed continental-scale surface temperature patterns and multi-decadal trends, including the more rapid warming since the mid-twentieth century, and the cooling immediately following large volcanic eruptions. The simulation of large-scale patterns of precipitation has been less successful and models perform less well for rainfall than for surface temperatures. The ability to simulate ocean thermal expansion, changes in glacier mass and ice sheets, and thus sea-level, has improved since the previous IPPC assessment report in 2007, but difficulties remain in accurately modelling the dynamics of the Greenland and Antarctic ice sheets. However, recent improvements and advances in scientific understanding have resulted in more reliable sea-level projections (UNEP, 2015).

The projections of future climate change are based on information described in scenarios of greenhouse gas and air pollutant emissions and land-use patterns. Scenarios are generated by a range of approaches, from simple idealized experiments to the more complex 'idealized assessment models' (IAMs). The key factors incorporated into the models are economic and population growth, lifestyle and behavioural changes, changes in energy consumption and land use, new technology and climate policy. The standard scenarios used in the IPCC's 5th Assessment Report are called Representative Concentration Pathways or RCPs.

In the IPCC scenarios, there are four RCPs, each projecting a different pathway of greenhouse gas and air pollutant emissions into the final decades of the twenty-first century. They include a strict mitigation scenario, RCP2.6, two intermediate scenarios, RCP4.5 and RCP6.0, and one business-as usual scenario with continuing high GHG emissions: RCP8.5. RCP2.6 represents a scenario that aims to keep global warming to not more than 2 °C above pre-industrial era temperatures. This scenario requires substantial global reductions in GHG emissions to take place in the first quarter of the twenty-first century, and net negative GHG emissions by 2100 – meaning that more greenhouse gases are sequestered than released into the environment, but this can only happen if GHG emissions are driven down to very low levels.

The business-as-usual scenario, RCP8.5, leads to global temperature increases of more than 4 °C by the end of the century – a situation that most experts consider will be close to catastrophic.

Which of the four RCP scenarios is the more likely? The RCP8.5 scenario is close to a business-as-usual scenario and that now seems unlikely, given the substantial international effort now underway to reduce GHG emissions and to mitigate the impact of climate change. The minimum impact scenario, RCP2.6, seems overly optimistic. It is based on the assumption that climate mitigation measures are widely implemented, and that they start to have a measurable impact within the next few years and definitely before 2020.

The most probable concentration pathway between now and the end of the twenty-first century is therefore likely to be represented by one of the intermediate scenarios: RCP4.5 or RCP6.0, or somewhere in between. These RCP scenarios lead to increased surface temperatures of between 2 and 3 °C before the end of the century.

But higher mean global temperatures are certainly not impossible. In the absence of near-term and much stronger mitigation actions and further commitments to reduce emissions, analyses in 2014 suggested that the likelihood of 4 °C warming being reached or exceeded this century had increased. According to a World Bank study at the time, there was about a 40% chance of exceeding 4 °C by 2100, and a 10% chance of exceeding 5 °C (World Bank, 2014).

The 2015 climate change agreement negotiated in Paris by over 180 countries was rightly celebrated as an historic agreement. The aim is to keep global warming to under 2 °C, although the governments of many small island states argued persuasively that even 2 °C of warming is too high: they insisted that an increase of 1.5 °C should be the absolute limit.

Unfortunately, it soon became apparent that even the 2 °C target was unlikely to be met. In May 2016, only a few months after the historic signing of the agreement, the UNFCCC acknowledged that CO_2 emissions would continue to rise until at least 2030:

> *If only the unconditional components of the INDCs are taken into account, global total emissions are projected to be 55.6 (53.1 to 57.3) $GtCO_2e$ in 2025 and 57.9 (54.4 to 59.3) $GtCO_2e$ in 2030, while including the conditional components of the INDCs lowers the estimated levels of such emissions to 54.1 (51.4 to 55.8) $GtCO_2e$ in 2025 and 55.5 (52.0 to 57.0) $GtCO_2e$ in 2030.*

In other words, even under the most optimistic scenario where substantial international funding enables all the conditional-based action to be implemented and achieved, CO_2 emissions will continue to increase over the next 15 years with no sign that they will even level out (UNFCCC, 2016).

Disaster risk management in Cuba

Cuba has made disaster risk management a high priority and set up an effective preparedness system. In 1963, Hurricane Flora caused 1200 fatalities; 25 years later, Hurricane George, just as strong, killed just four people in Cuba as opposed to 600 people in other countries in the region. In 2008, Hurricanes Ike and Gustav, which together claimed almost 350 lives in other countries in the Caribbean, killed only seven people in Cuba.

The Cuban government has addressed the problem of disaster management at all levels. The Civil Defence System has evolved, and in 2005 the Joint Staff of National Defence (EMNDC), with the support of the UNDP, founded Risk Reduction Management Centres (RRMCs). These centres are responsible for conducting research, collecting data, checking vulnerability, disseminating information, coordination and preparedness. Education on disaster risk management has integrated awareness and preparedness into the social fabric of society. As a result, people are in a position to analyze the data available individually to gauge the potential threat. Coordination and constant citizen engagement are the twin pillars of successful disaster risk management in Cuba.

In 2012, Hurricane Sandy destroyed more than 300 000 homes and affected more than 3 million people in Cuba. Despite the strength of the storm, Hurricane Sandy claimed only 11 lives in Cuba, while other countries suffered a total death toll of 285.

Source: UNDESA (2014).

Warming the Oceans

Most of the discussion about climate change has focused on the warming of the atmosphere – certainly this was the focus of attention at the December 2015 meeting in Paris. But the warming of the oceans is actually the real problem, and one that has only recently come into serious focus. A detailed report issued by the IUCN in 2016 examined the consequences of ocean warming in considerable detail (Laffoley & Baxter, 2016).

The oceans of water on planet Earth hold truly enormous quantities of thermal energy: on a volumetric basis, the heat capacity of water is roughly 4000 times that of air. What this means is that the oceans are an immense heat sink. Over 90% of the increased thermal energy generated by the Earth's present energy imbalance is estimated to have been absorbed by the oceans. This enormous heat sink has huge thermal inertia, effectively buffering the atmosphere from rapid fluctuations in temperature. The constant movement of ocean water also distributes heat around the planet, transporting heat away from the tropics towards the poles. But even though the water temperatures are increasing extremely slowly, they can be measured and correlated. The trend is clear. The effects of ocean warming on the global environment are summarized by the IUCN in Table 1.1.

Table 1.1 Ocean warming: Its impact and consequences.

Ocean warming effect	Probable impacts
Changes in ocean heat content (OHC)	• Increasing uptake of heat by the ocean as a response to the Earth's energy imbalance buffering atmospheric warming • Rising water temperatures at all depths • Intensification of El Nino events
Warming of adjacent land masses	• Warmer land surface temperatures • Melting permafrost • Retreating mountain glaciers and surface melting of the Greenland ice sheet • Terrestrial vegetation changes • Increased extent and magnitude of forest fires
Rising sea-levels due to expansion of water with temperature plus melting ice sheets and glaciers	• Permanent inundation of coastal areas and low-elevation islands and atolls • Increased coastal flooding from storm surges • Saltwater intrusion into aquifers
Melting cryosphere (the frozen world)	• Basal melting and thinning of ice shelves in Antarctica destabilizing dammed-up ice sheet glaciers • Increased overall Antarctic ice mass and sea ice • Accelerated mass loss of the Greenland Ice Sheet including basal melting of marine terminating glaciers • Accelerating reduction of sea ice in the Arctic and in the Bellinghausen/Amundsen Sea of Antarctica
Intensification of the hydrological cycle	• Enhanced atmospheric moisture transport towards the poles • Rising humidity and increasing precipitation • Increased Eurasian river discharges • Extreme droughts and floods • Both negative and positive feedbacks to global warming • Salinity increasing where evaporation dominates in the mid latitudes and decreasing in the rainfall-dominated regions of the tropical and polar seas
Negative feedback on the ocean carbon sink	• Higher sea surface temperature reduces CO_2 uptake from the atmosphere, increasing the rate of increase of atmospheric CO_2
Deoxygenation	• Reduced oxygen solubility in warmer water • Reduced penetration of oxygen into deeper water due to enhanced stratification
Acidification	• Rising temperatures reinforces ocean acidification
More extremes in natural variability such as the El Nino/Southern Oscillation (ENSO) and in weather events	• Changes in the occurrence, frequency and severity of cyclones/hurricanes • Changes in the location and meandering of jet streams affecting downstream weather • Landslides, collapses in fisheries • Coral bleaching, enhanced disease prevalence, malnutrition and human migration • Monsoons, forest fires and associated air pollution
Changes in biological processes at cellular to ecosystem scales	• Reduction and possible collapse of fisheries • Increasing food insecurity

Source: Adapted from Laffoley & Baxter (2016). Reproduced with permission of the International Union for Conservation of Nature.

Multidimensional Threats

There are essentially ten different types of threat caused or exacerbated by climate change.

1) **Sea-level rise and coastal flooding.** Populations, economic activity and infrastructure in low-lying coastal zones are vulnerable. Urban populations are frequently unprotected due to substandard housing and inadequate insurance. Marginalized rural communities with multidimensional poverty and limited alternative livelihoods and coping mechanisms are particularly at risk. Rising sea-levels contribute to seawater intrusion into freshwater aquifers. Island populations are extremely vulnerable to this threat.

2) **Extreme precipitation and inland flooding.** Large numbers of people in urban areas are exposed to flood events, particularly in low-income informal settlements and shanty towns. Poorly maintained and inadequate urban drainage infrastructure may have limited ability to cope.

3) **Systemic failures.** Populations and infrastructure exposed and lack historical experience with hazards such as electrical distribution system failures, communication systems failures, health and emergency response systems collapse.

4) **Increasing frequency and intensity of extreme heat, including heat island effects.** Extreme heat events are expected to increase in frequency and to impact a larger area of land. Urban populations of the elderly, the very young, expectant mothers, and people with chronic health problems in settlements are incapacitated by higher temperatures. Local organizations that provide health, emergency, and social services are unable to adapt effectively to new risk levels for vulnerable groups. Heatwaves are now a growing global threat, killing thousands of people each year (*The Lancet*, 2015b). In the US, a 2016 estimate of deaths from extreme events over the period 2004–2013 showed that heatwaves cause more fatalities than more spectacular events such tornadoes and hurricanes (USGCRP, 2016).

5) **Warming, drought and precipitation variability.** Changes in precipitation will occur with continued warming, with substantial adverse consequences for the availability of water in many regions. Poorer communities in urban and rural settings are increasingly susceptible to food insecurity, particularly subsistence farmers and people in low-income, agriculturally dependent economies. There may be limited ability to cope among the elderly and female-headed households. Terrestrial ecosystems will be stressed and less productive; the services they provide to local communities will be disrupted.

6) **Drought.** Crop yields are already declining in many regions. Reduced yields and production losses increase rapidly above 1.5–2 °C warming. Urban populations with inadequate water services, existing water shortages and supply problems are likely to be severely impacted. Poor farmers in dry lands or pastoralists with insufficient access to drinking and irrigation water will suffer. Limited ability to compensate for losses in water-dependent farming and pastoralist systems will engender conflict.

7) **Rising ocean temperatures and acidification.** Warm-water coral reefs are highly susceptible to both increased temperatures and higher levels of acidity. Marine ecosystems on which coastal communities depend will deteriorate. Invasive marine species will proliferate.

8) **Wildfires.** Although not usually on the list of threats driven by climate change, there is increasing evidence that wildfires, forest fires and bush fires are increasing in extent, frequency and ferocity. In 2015, the worst year on record, there were over

100,000 wildfires worldwide, including massive forest fires in Indonesia that could clearly be seen from space, and which generated so much smoke and air pollution across Indonesia, Singapore and Malaysia that the health impact on those populations is certain to be significant – not to mention the impact on wildlife in the forests that are destroyed (De Groot, 2015; GFED, 2015).

9) **Societal upheaval and strife.** There is clear evidence that climate change is already affecting livelihoods in many regions of the world. Climate change has a greater adverse impact on poorer communities that are less able to cope and adapt. Coastal zones will become increasingly threatened by inundation and extreme weather, while at the same time livelihoods that depend on agriculture and fishing will become more difficult and less viable. Many communities will be forced to move and conflict between social groups will be inevitable. The competition for resources, particularly water, will become sharper. Climate-change driven migration is increasingly becoming an international issue (*The Lancet*, 2015b).

10) **Health.** The health impacts of climate change have only recently been coming into focus. They include malnutrition, impacts on mental health, increased levels of cardiovascular disease, respiratory disease, and vector-borne diseases (*The Lancet*, 2015a).

The impacts of climate change on health are summarized in Table 1.2. This analysis is from a report that focused on the US, but the consequences of the changing climate on human health are global (USGCRP, 2016).

To bring this global overview down to the level of the islands themselves, Table 1.3 sets out the expected impacts of climate change on the British Virgin Islands in the Caribbean (BVI, 2011).

This is all seriously bad news for small islands. Increased global temperatures means higher sea-levels, more intense storms and extreme weather, coupled with the probability of intermittent droughts and the certainty of seriously degraded marine resources. Several small islands will almost certainly become uninhabitable before the end of the century. Nearly all islands, even the larger ones, will suffer severe, damaging, and lasting social, economic and environmental impacts. These issues are discussed in greater detail in the chapters that follow.

Climate change and the price of food

Agriculture is one of the most important economic sectors in many poor countries. It is also critical for household food security. In fact, some of the most severe poverty impacts of climate change are expected to be channelled through agriculture.

Although climate change impacts on crop yields and prices are uncertain and hard to predict, the IPCC has concluded with high confidence that, in the longer term, crop production will be consistently and negatively affected by climate change, particularly in low-latitude countries (where all the SIDS are located).

Poor people are highly vulnerable to food price hikes. The higher share of income that poor people spend on food makes them particularly impacted by rising prices or price volatility on food items. In the developing world, the poorest households spend 40–60% of their income on food and beverages compared with less than 25% for wealthier households. In spite of the benefits to farmers who are net sellers of food, studies tend to agree that in the absence of changes in production and wages, a rise in food prices will increase poverty in most countries due to the negative impact on consumers. In a sample of

28 developing countries, the global price spikes between June and December 2010, which increased food prices by an average of 37%, increased the number of people in extreme poverty by 44 million. The 2008 food price shock, which resulted in price increases of over 100%, resulted in an additional 100 million people in extreme poverty. In Haiti in April 2008, there were riots and violent conflict with the civil authorities that eventually led to the downfall of the government.

Source: Hallegatte *et al.* (2016).

Table 1.2 Health impacts of climate change.

Threats	Climate change	Exposure	Health outcome	Impact
Extreme heat	More frequent, severe and prolonged heat events	Elevated temperature	Heat-related death and illness	Rising temperatures will lead to an increase in heat-related deaths and illnesses
Outdoor air quality	Increasing temperatures and changing precipitation patterns	Worsened air quality (ozone, particulates, and higher pollen counts)	Premature death, acute and chronic cardiovascular and respiratory illnesses	Rising temperatures and wildfires are decreasing precipitation which will lead to increases in ozone and particulate matter, elevating the risks of cardiovascular and respiratory illnesses and death
Flooding	Rising sea-levels and more frequent or intense extreme precipitation, hurricanes, and storm surge events	Contaminated water, debris and disruptions to essential infrastructure	Drowning, injuries, mental health consequences, gastrointestinal and other illnesses	Increased coastal and inland flooding exposes populations to a range of negative health impacts before, during and after events
Water-borne infections (Lyme disease)	Changes in temperature extremes and seasonal weather patterns	Earlier and geographically expanded tick activity	Lyme disease	Ticks will show earlier seasonal activity and a generally northward range expansion, increasing risk of human exposure to Lyme-disease carrying bacteria
Water-related infection (*Vibrio unifaces*)	Rising sea surface temperatures, changes in precipitation and runoff affecting coastal salinity	Recreational water or shellfish contaminated with *Vibrio vulnificus*	*Vibrio vulnificus* induced diarrhoea and intestinal illness, wound and blood-stream infections, death	Increases in water temperatures will alter timing and location of *Vibrio vulnificus* growth, increasing exposure and risk of water-borne illness
Food-related infection (*Salmonella*)	Increases in temperature, humidity, and season length	Increased growth of pathogens, seasonal shifts in incidence of *Salmonella* exposure	*Salmonella* infection, gastrointestinal outbreaks	Rising temperatures increase *Salmonella* prevalence in food; longer seasons and warming winters increase risk of exposure and infection

Table 1.3 Climate change impacts in the British Virgin Islands.

Impact areas	Potential and existing climate change impacts
Beach and shoreline stability	• Increase in beach erosion and shrinkage • Shoreline retreating and more vulnerable to flooding
Coastal and marine ecosystems	• Coral reefs experiencing increased bleaching, structural damage, disease and death • Landward migration or inundation of mangroves and increased mortality • Decreased growth of seagrass beds and increased stress and mortality
Critical infrastructure	• Road network, critical facilities, utilities, developable lands and the sewage system (especially coastal) all at great risk of damage
Human settlements	• Homes and developable land (especially those in the coastal zone) at greater risk of damage
Energy security	• Energy generation and distribution system at greater risk of damage • Increase in energy costs, increase in energy use for cooking
Food security and agriculture	• Decrease in agricultural yield (or increased cost of production) due to decrease in rainwater • Increase in agricultural pests, weeds, diseases and invasive species • Decrease in agricultural produce (or increase in cost). Less rainwater for agriculture • Soil degradation, resulting in reduced yield • Increase in crop damage and disruption of production cycles • Increased stress to livestock, resulting in decreased productivity • Changes to imported food availability, increased cost and/or poorer quality

References

BVI (2011) *The Virgin Islands' Climate Change Policy: Achieving Low-carbon, Climate Resilient Development*. Government of the Virgin Islands. Available at www.car-spaw-rac.org/IMG/pdf/Climate_Change_Adaptation_Policy_-_Virgin_Islands.pdf

De Groot, W.J., Flannigan, M.D., and Stocks, B.J. (2015) *Climate Change and Wildfires*. In Proceedings of the 4th International Symposium on Fire Economics, Planning and Policy: Climate Change and Wildfires. US Department of Agriculture, Forest Service, Pacific Southwest Research Station, Albany, CA, pp. 1–10.

GFED (2015) Global Fire Emissions Database. Available at https://climatedataguide.ucar.edu/climate-data/gfed-global-fire-emissions-database

Hallegatte, S., Bangalore, M., Bonzanigo, L., *et al.* (2016) *Shock Waves: Managing the Impacts of Climate Change on Poverty*. Climate Change and Poverty Series. World Bank, Washington, DC.

IPCC (2014a) *Climate Change 2014: Synthesis Report. Contribution of Working Groups I, II and III to the Fifth Assessment Report of the Intergovernmental Panel on Climate Change*. IPCC, Geneva.

IPCC (2014b) Technical summary, in *Climate Change 2014: Impacts, Adaptation, and Vulnerability. Part A: Global and Sectoral Aspects. Contribution of Working Group II to* the Fifth Assessment Report of the Intergovernmental Panel on Climate Change (eds C.B. Field, V.R. Barros, D.J. Dokken, *et al.*). Cambridge University Press, Cambridge, pp. 35–94.

Laffoley, D. and Baxter, J.M. (eds) (2016) *Explaining Ocean Warming: Causes, Scale, Effects and Consequences*. International Union for the Conservation of Nature, Gland, Switzerland.

LeQueré *et al.* (2016) Global Carbon Budget 2016. *Earth Syst. Sci. Data*, 8, 605–649. doi:10.5194/essd-8-605-2016. Available at www.earth-syst-sci-data.net/8/605/2016/

Meadows, D.H., Meadows, D.L., Randers J., and Behrens III, W.W. (1972) *The Limits to Growth: A Report of the Club of Rome's Project on the Predicament of Mankind*. Universe Books, New York.

NASA (2015a) NASA, NOAA find 2014 warmest year in modern record. Available at www.giss.nasa.gov/research/news/20150116

NASA (2015b) Global annual mean surface air temperature change. Available at http://data.giss.nasa.gov/gistemp/graphs/

NASA (2015c) Globally averaged sea level change. Available at www.giss.nasa.gov/research/features/201508_risingseas/

NOAA (2015) 2013 State of the Climate: Carbon dioxide tops 400 ppm. Available at www.climate.gov/news-features/understanding-climate/2013-state-climate-carbon-dioxide-tops-400-ppm

The Lancet (2015a) *Health and climate change: policy responses to protect public health*. The Lancet Commissions. doi:10.1016/S0140-6736(15)60854-6

The Lancet (2015b) Adapting to migration as a planetary force. Editorial published online. *The Lancet*, 386, September 12.

UNDESA (2014) *Trends in Sustainable Development: Small Island Developing States*. Department of Economic and Social Affairs, United Nations, New York.

UNEP (2015) *The Emissions Gap Report 2014*. United Nations Environment Program, Nairobi.

UNEP (2016) *The Emissions Gap Report 2015*. United Nations Environment Program, Nairobi.

UNFCCC (2016) Aggregate effect of the intended nationally determined contributions: an update. Synthesis report by the secretariat. United Nations Framework Convention on Climate Change. Report FCCC/CP/2016/2.

USGCRP (2016) *The Impacts of Climate Change on Human Health in the United States: A Scientific Assessment*. US Global Change Research Program, Washington, DC.

WMO (2017) *WMO Statement on the State of the Global Climate in 2016*. WMO No. 1189. World Meteorological Organization, Geneva.

World Bank (2014) *Turn Down the Heat, Confronting the New Climate Normal*. World Bank, Washington, DC.

2

Small Island Developing States

Islands are particularly vulnerable to the effects of climate change. Towns and villages tend to cluster on the coast and are therefore exposed to greater risk from sea-level rise, storm surges, and coastal zone flooding. On many islands the capital city and government ministries are also on the coast and risk being damaged and dysfunctional at precisely the time that their disaster management capability and organizational competence are most required. As the days get hotter, periods of drought will become longer and more frequent. At the same time, storms are likely to become more intense, with flooding and landslides occurring more often. Where fishing is an important livelihood for a significant part of the population, the effects of sea-level rise and the acidification of the oceans are almost certain to disrupt and endanger fishers' livelihoods. Smaller islands are more vulnerable still; many are low-lying and at risk from being completely swamped by storm surges. Several island states may disappear entirely over the next 20 years as sea-levels rise – their populations forced to emigrate to higher ground in neighbouring countries better able to withstand storm events and higher sea-levels. The social and economic dislocation, and potentially the conflict, caused by these mass movements of island populations will have serious political and economic regional consequences.

The special case of small island developing states (SIDS) was first recognized by the international community at the UN Conference on Environment and Development (UNCED) held in Rio de Janeiro in 1992. The focus at that time was on sustainable development. In Agenda 21 it was stressed that:

> Small island developing States and islands supporting small communities are a special case both for environment and development. They are ecologically fragile and vulnerable. Their small size, limited resources, geographic dispersion, and isolation from markets, place them at a disadvantage economically and prevent economies of scale.

Since then, several international frameworks have been established that refer to the constraints faced by SIDS in achieving sustainable development. In 1994, the Barbados Programme of Action (BPOA) transformed the recommendations made by Agenda 21 into concrete actions and measures intended to enable SIDS to achieve sustainable development. In 2005, the Mauritius Strategy for the further implementation of the BPOA was adopted with a view to addressing the implementation gap that continued to confront SIDS at that time. In 2010, a high-level meeting was held during the 65th session of the UN General Assembly to carry out a five-year review of the progress made in addressing the vulnerabilities of SIDS through the Mauritius Strategy.

During the UN Conference on Sustainable Development held in Rio de Janeiro in June 2012, the conference noted once again the particular difficulties faced by small island developing states. The conference called for a third international conference on SIDS, a call that was answered by Samoa, which offered to host the conference in 2014. The conference was held in Apia, Samoa, in September 2014, and led to the declaration of the 'SAMOA Pathway'. The name stands for Small Island Development States Accelerated Modalities of Action. Climate change was identified as one of the priority actions, with the SIDS requesting assistance from the international community to enable them to:

- Build resilience to the impacts of climate change and to improve their adaptive capacity through the design and implementation of climate change adaptation measures appropriate to their respective vulnerabilities and economic, environmental, and social situations;
- Improve the baseline monitoring of island systems and the downscaling of climate model projections to enable better projections of the future impacts on small islands;
- Raise awareness and communicate climate change risks including through public dialogue with local communities, to increase human and environmental resilience to the longer-term impacts of climate change;
- Address remaining gaps in capacity for gaining access to and managing climate finance.

However, what is now clear is that development can never be sustainable if countries are vulnerable to the impacts of climate change. Small islands suffer substantial economic losses when impacted by extreme weather linked to climate change. In poorer countries, governments typically lack the resources to support communities trying to cope with the damage wrought by storms and flooding. Families are forced to sell possessions and physical capital in order to raise money to repair housing and pay for healthcare for the injured. Communities become poorer and even more vulnerable. If extreme weather events become more frequent and more damaging, many families never recover and descend permanently into poverty.

Meet the SIDS

The United Nations recognizes 51 Small Island Developing States (UN, 2011).[1] They are generally grouped by geographical region, as shown in Table 2.1.

Not all the SIDS are islands: Belize, Guyana, and Suriname are also included. Singapore seems incongruous, given that it is definitely not a developing country. However, the republic includes several dozen smaller islands that are vulnerable to climate change impacts and this presumably explains its inclusion in the group.

Table 2.2 provides some basic data on the 51 states (UNDESA, 2016). It is a diverse group, ranging from countries larger than Germany (Papua New Guinea has an area of

---

1 This 2011 UN report lists 52 SIDS including the Netherlands Antilles. However, this group of islands was dissolved in 2010 to become Aruba, Bonaire, Curacao and several smaller islands. In this book, the Netherlands Antilles is not counted as a SIDS.

Table 2.1 The geographical regions of the 51 Small Island Developing States.

Pacific	American Samoa	Cook Islands	Fiji	French Polynesia
	Guam	Kiribati	Marshall Islands	Micronesia, states
	Nauru	New Caledonia	N. Mariana Islands	Niue
	Palau	Papua New Guinea	Samoa	Solomon Islands
	Timor-Leste	Tonga	Tuvalu	Vanuatu
Caribbean	Anguilla	Antigua & Barbuda	Aruba	Bahamas
	Barbados	Belize	British Virgin Islands	Cuba
	Dominica	Dominican Republic	Grenada	Guyana
	Haiti	Jamaica	Montserrat	Puerto Rico
	St Kitts and Nevis	St Lucia	St Vincent & Grenadines	Suriname
	Trinidad and Tobago	US Virgin Islands		
AIMS [*]	Bahrain	Cabo Verde	Comoros	Guinea Bissau
	Maldives	Mauritius	Sao Tome and Principe	Seychelles
	Singapore			

[*] AIMS covers Africa, Indian Ocean, Mediterranean and South China Sea.

462,840 km^2) down to tiny coral islands like Tuvalu, which are smaller than an average European airport.

The wealthiest island states, using per capita GNP as the metric, are Singapore, the British Virgin Islands, and New Caledonia – all with per capita GNP over $30,000. The poorest are Haiti, Guinea-Bissau and Comoros, with per capita GNP less than $1000. Although the SIDS countries are characterized as 'developing', only nine of them are in fact classified as LDCs (less developed countries) under the UN system: Comoros, Guinea-Bissau, Sao Tome and Principe, Timor-Leste, Kiribati, Solomon Islands, Tuvalu, Vanuatu and Haiti (UN, 2015).

Unsurprisingly, the country with the highest population density is Singapore, which has almost 8000 people per km^2. The Maldives is in second place with 1233 people/km^2, followed by Barbados and Mauritius with 663 and 649 persons/km^2 respectively.

In terms of emissions of CO_2, the states emitting more than 30 MtCO_2 per year are Bahrain, Cuba, Singapore, and Trinidad and Tobago – all countries with petroleum refineries and natural gas distribution systems. At the other end of the scale are island states that emit almost negligible quantities of CO_2: more than half the island states emit less than 1 MtCO_2 annually, which is just 0.003% of global emissions.

It is important to note that emissions of carbon dioxide are extremely small for these states. The *total* contribution of *all* the SIDS to atmospheric CO_2 emissions is minimal – less than 1% of the global total.[2]

---

2 Total CO_2 emissions from the SIDS listed in Table 2.2 are 0.234 GtCO_2. Global CO_2 emissions in 2014 were estimated as 35.5 GtCO_2 (UNEP, 2015). Although CO_2 emission data for American Samoa, Guam, the Northern Mariana Islands, Tuvalu, and the US Virgin Islands are unavailable, they are unlikely to push the total above 1% of global emissions for the full set of SIDS.

Table 2.2 Basic data on the 51 SIDS.

	Small Island Developing State	Population (2014) 000s	Surface area km²	GNP per capita (2014) USD	CO₂ emissions (2013) ktCO₂
1	American Samoa	56	199	*	*
2	Anguilla	15	91	21493	136
3	Antigua and Barbuda	83	442	13731	524
4	Aruba	104	180	25751	876
5	Bahamas	393	13940	22218	3110
6	Bahrain	1397	771	24854	31958
7	Barbados	285	430	15360	1448
8	Belize	367	22966	4831	517
9	British Virgin Islands	31	151	30502	176
10	Cabo Verde	527	4033	3609	444
11	Comoros	807	2235	841	161
12	Cook Islands	21	236	15003	70
13	Cuba	11393	109884	7274	39340
14	Dominica	73	750	7361	132
15	Dominican Republic	10649	48231	6147	22072
16	Fiji	898	18272	5112	1709
17	French Polynesia	286	4000	20099	821
18	Grenada	107	345	8313	304
19	Guam	172	549	*	*
20	Guinea-Bissau	1888	36125	672	257
21	Guyana	771	214969	4040	1936
22	Haiti	10848	27750	813	2406
23	Jamaica	2803	10991	5004	7726
24	Kiribati	114	726	1632	62
25	Maldives	370	300	8484	1049
26	Marshall Islands	53	181	3948	103
27	Mauritius	1278	1969	9945	3726
28	Micronesia Fed. States	105	702	2960	147
29	Montserrat	5	103	12384	51
30	Nauru	10	21	17857	44
31	New Caledonia	266	18575	39392	3861
32	Niue	2	260	*	11
33	Northern Mariana Islands	55	457	*	*

Table 2.2 (Continued)

Small Island Developing State	Population (2014) 000s	Surface area km²	GNP per capita (2014) USD	CO$_2$ emissions (2013) ktCO$_2$
34 Palau	22	459	11068	224
35 Papua New Guinea	7776	462840	2221	6073
36 Puerto Rico	3681	8868	28123	*
37 Saint Kitts & Nevis	56	261	15510	279
38 Saint Lucia	186	539	7655	407
39 Saint Vincent & Grenadines	110	389	6669	209
40 Samoa	195	2842	4294	238
41 Sao Tome & Principe	194	964	1811	114
42 Seychelles	97	457	15759	645
43 Singapore	5697	718	55910	50557
44 Solomon Islands	595	28896	1927	198
45 Suriname	548	163820	9680	2101
46 Timor-Leste	1211	14919	4294	440
47 Tonga	107	747	4122	209
48 Trinidad & Tobago	1365	5130	20723	46542
49 Tuvalu	10	26	3796	*
50 US Virgin Islands	106	347	*	*
51 Vanuatu	271	12189	3138	106

Demography

By definition, the SIDS have relatively small populations. The lowest populations are in the Pacific region: the 20 Pacific SIDS average just over half a million people, and if Papua New Guinea is set aside, the mean falls to about 200,000. However, many islands are small in area, so population densities can be very high. Apart from Singapore, which is consistently a statistical outlier, the highest population densities are found in the Maldives.

In all regions, population growth is slowing – from about 2% over 1975–1980 to 1.3% over the period 2005–2010 – and it is projected to fall further (UNDESA, 2014). In some island states, emigration has contributed significantly to a further decline. In Micronesia, population growth is negative, and in Polynesia the rate of increase falls from 1.6 to 0.6% when emigration is factored in. People emigrate for a variety of reasons, but it is reasonable to suppose that many people leave because of the risk of natural disasters and economic vulnerability.

In general, life expectancy is increasing in the island states, rising from about 65 years in 1990 to over 70 years in 2013, and is projected to continue to increase over the next 20 years. This transition will change the age structure of SIDS populations: the fraction of the populations that is over 60 will gradually increase. These demographic changes have implications for the level of public services that will be required in the future, particularly in regard to health services – older populations are more vulnerable to the health impacts expected to result from climate change.

Social Development

The Human Development Indices of 35 SIDS are ranked in Table 2.3. Papua New Guinea, the Solomon Islands, Comoros, Haiti and Guinea-Bissau have the lowest ranking among the SIDS (HDR, 2014).

The Human Development Index is important because it reflects the level of government services provided to a country's population. A lower index indicates that the majority of the population lack essential services related to healthcare and education. These factors have implications for the ability of a country's population to cope with and recover from shocks. The mean value for the world is 0.711; 14 of the SIDS shown in this table fall below the global mean.

Is the quality of life improving in the island states? The HDR report shows the trend lines for the countries listed in Table 2.2. All the countries show improvements in their human development indices except for two: for Belize and Timor-Leste the trend is downwards.

Gender equality and employment

SIDS have been making progress in terms of gender equality in recent decades; an increasing number of women are involved in government and other elements of public service that were once dominated by men. The number of women representatives in national parliaments is rising: over the last decade the number of seats occupied by women has increased by about 5%, although as a group the SIDS still lag behind the global average.

Like many other regions of the world, the number of women in the labour force has been increasing. Among the SIDS regions, the Caribbean has seen the greatest change compared with the AIMS and Pacific regions. In education, the SIDS have made progress; in all regions the years of schooling among children has increased significantly. Enrolment in primary education has also increased, and primary school attendance is relatively high in all three regions.

Employment opportunities are often scarce in island economies, and the rate of unemployment among youth is higher than among adults. In 2011 the rate of youth unemployment was over 35%, the highest levels being in the Caribbean region.

The level of crime in many small islands is above the global average. Data from 19 SIDS reported in 2011 showed an average homicide rate of 19 per 100,000 persons, three times more than the global mean.

SIDS are net importers of food and therefore vulnerable to variable availability and fluctuating prices.

Source: UNDESA (2014).

Economic Vulnerability

The UN Department of Economic and Social Affairs has developed a methodology to enable that agency to assess and characterize the economic vulnerability of many of the poorer countries, including many of the SIDS (UN, 2008). The Economic Vulnerability

Table 2.3 Human Development Indices for the SIDS.

Very high 0.8–1.0	High 0.7–0.8		Medium 0.55–0.8	Low <0.55
Singapore	Bahamas	Saint Lucia	Cabo Verde	Solomon Islands
Bahrain	Barbados	Grenada	Micronesia, Fed. St.	Papua New Guinea
	Antigua & Barbuda	Dominica	Guyana	Comoros
	Palau	Saint Vincent & Grenada	Timor-Leste	Haiti
	Mauritius	Jamaica	Vanuatu	Guinea-Bissau
	Seychelles	Dominican Republic	Kiribati	
	Trinidad & Tobago	Maldives	Sao Tome & Principe	
	Cuba	Samoa		
	Saint Kitts & Nevis	Seychelles		
	Fiji	Belize		
	Tonga			

Index, or EVI, captures the relative risk posed to a country's development by exogenous shocks. The impact of these shocks depends on their magnitude and frequency, and also on the structural characteristics of the country concerned, which affect the degree to which it is exposed to such shocks and its capacity to withstand them (i.e. its resilience). Given the potentially huge economic impacts of climate change and the shocks of extreme weather, a country's EVI index provides a useful metric of a country's resilience to climate change stressors.

The EVI is composed of seven indicators:

- Population size
- Remoteness
- Merchandise export concentration
- Share of agriculture, forestry and fisheries in GDP
- Homelessness due to natural disasters
- Instability of agricultural production
- Instability of exports of goods and services

The first four indicators contribute to an Exposure index; the last three indicators are combined to give a Shock index. The Exposure index and the Shock index are combined in equal measure to give the Economic Vulnerability Index. The relevance of the seven indicators to climate change impact is summarized below.

Population

Larger countries are generally more resilient to shocks and have a more diversified economy because of economies of scale supported by a relatively large domestic market.

Small size is often associated with a lack of structural diversification and a dependence on external markets. Smaller economies therefore have a higher exposure to natural shocks, and many small low-income countries are situated in regions that are prone to natural disasters. The size of a country's population is therefore considered to be a major indicator of economic vulnerability.

Remoteness
A country's location is a factor that has a bearing on exposure and resilience. Countries situated far from major world markets face a series of structural handicaps, such as high transportation costs and isolation, which makes them less able to respond to shocks in an effective way. Countries isolated from main markets have difficulty in diversifying their economies. Remoteness is a structural obstacle to trade and growth, and a possible source of vulnerability when shocks occur. It is considered to be one of the main handicaps of many low-income small island states.

Merchandise Export Concentration
Export concentration generally increases a country's exposure to trade shocks.

Share of Agriculture, Forestry and Fisheries in GDP
A larger share of agriculture, forestry and fisheries in GDP implies a higher exposure to shocks both in relation to terms of trade and to natural disasters.

Homelessness Due to Natural Disasters
The indicator provides information on the average share of the population that is displaced by natural disasters over a period of time.

Instability of Agricultural Production
This indicator measures the instability of agricultural production with respect to its trend line. The trend value reflects factors which may be permanent in nature (such as the availability of arable land) as well as economic policies, while fluctuations around the trend may capture the occurrence of natural shocks and their impact on production.

Instability of Exports of Goods and Services
For low-income countries, particularly those dependent on agricultural exports or tourism, instability of exports is a source of vulnerability. It results from fluctuations in world demand and other reasons not necessarily associated with the domestic policy of the country concerned, such as climate change or changes in policies of major importing markets.

The EVIs for 33 SIDS are shown in Figure 2.1, which compares EVI values with the less developed country (LDC) mean value of 45.7.

The least vulnerable countries in this group are Barbados and the Dominican Republic; the most vulnerable islands are Kiribati, Suriname, Tuvalu, Tonga and Guinea-Bissau. Almost half of the group, 15 SIDS, have an EVI value above the LDC mean. In terms of regional values, the Pacific region has greater vulnerability. Both the AIMS and Caribbean regions are just below the LDC mean EVI, rated at 44.0 and 42.3 respectively.

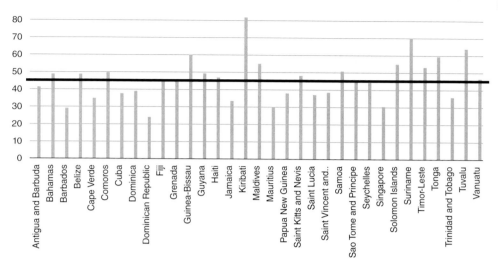

Figure 2.1 Economic vulnerability of 33 SIDS. *Source:* Data from www.un.org/en/development/desa/policy/cdp/ldc/excel/2012_ldc_data.xls

The Pacific region has an EVI of 55.0 – 20% higher than the LDC mean value. But many SIDS are missing from the dataset; it is likely that more of the islands in the Pacific would be found to be economically vulnerable if the data were available (UN, 2012).

Climate Change Impacts on Small Islands

The threats posed by the changing climate are well established: hotter days and nights for longer periods of time, rising sea-levels, warmer oceans and more acidic seawater, unpredictable and less precipitation interspersed with more intense and extreme weather events. All countries and communities will have to learn how to cope with these threats. In general, small islands will have a harder time adapting to, and coping with, the threats posed by climate change. The reasons for this can be summarized as follows:

- Small population
- Low elevation coastal zones
- Limited resources
- Remoteness
- Exposure to natural disasters
- Dependence on agriculture, forestry, and fishing

These factors are almost identical to the indicators identified by the UNDESA as defining the multidimensional characteristics of economic vulnerability. One aspect, though, makes a considerable difference: this is the exceptional vulnerability of many SIDS because of the extent of their low elevation coastal zones.

Coastal Zones: A Clear and Present Danger

The majority of small island states have low and flat coastal areas where a significant fraction of the population is resident, and where important infrastructure is located. Coastal communities on small islands are especially vulnerable to storms and extreme weather. Even events that occur hundreds of kilometres away, like tsunamis, can have a devastating impact on coastal communities. In short, the main threat to small islands comes from the sea, and the place where islands are most exposed is on their coastlines. Table 2.4 outlines the main threats to the coastal zones of small islands (Wong *et al.*, 2014).

Sea-level Rise

As noted in Chapter 1, measurements indicate that global mean sea-level rose by about 0.2 metres between 1901 and 2010, or about 2 mm a year. In the twenty-first century, the average rate has risen to more than 3 mm a year. However, there are large regional variations: since 1993 the regional rates for the western Pacific are up to

Table 2.4 Threats to coastal zones of small islands.

Threat	Physical and ecosystem impacts	Projections
Sea-level rise	Submergence, flood damage, saltwater intrusion, rising water tables, impeded drainage, damage to and loss of wetlands	Global sea-levels rising between 2 and 3 mm per year. Large regional short-term variations. Some islands in the Pacific may experience much larger rises
Storms: tropical cyclones	Storm surges and waves, coastal flooding, erosion, saltwater intrusion, impeded drainage, loss of wetlands, damage to coastal infrastructure and ecosystems	Frequency of tropical storms may be unchanged, but the intensity is projected to increase and therefore the severity of their impact on coastal systems
Winds	Wind-driven waves, storm surges, coastal currents, damage to infrastructure and ecosystems	Mean wind speeds are not forecast to increase but tropical cyclones are expected to increase in intensity
Waves	Coastal erosion, overtopping, and coastal flooding	Likely increase in higher latitudes; possible increase in southern oceans
Sea surface temperature	Changes to stratification and circulation patterns. Increased coral bleaching and mortality, polewards species migrations, increased algae blooms	Sea surface temperatures (SSTs) are rising globally. Coastal SSTs are rising faster than in the open ocean.
Freshwater input	Altered flood risk in coastal lowlands, altered water quality and salinity, altered fluvial sediment supply, altered circulation and nutrient supply	Probably net declining trend, but possible increases in higher latitudes and wet tropics
Ocean acidity	Increased CO_2 fertilization, increased pH (acidity), and carbonate ion concentration	Acidity is increasing at unprecedented rates but with local and regional variations

three times larger than the global mean, while those for much of the eastern Pacific are near zero or negative.

Sea-level rise (SLR) is not uniform across the globe. Normal climate variations in different regions affect both interannual and interdecadal trends. Coastlines near glaciers and ice sheets may actually show a declining trend because land tends to rise as ice sheets melt, and the reduced land mass also changes the shape of the sea floor. In the tropical western Pacific, where a large number of small island communities are located, rates of sea-level rise of up to four times the global average (approximately 12 mm per year) have been reported between 1993–2009 (Nurse *et al.*, 2014). In the Pacific region the ENSO plays a strong role in regional sea-levels, with lower than average sea-level during El Nino events and higher than average levels during La Nina events. Large variations have also been recorded in the Caribbean, but the observed average rate of SLR over the last 60 years is similar to the global average.

Man-made factors that influence local sea-level include sediment consolidation from building loads, reduced sediment delivery to the coast, and the extraction of subsurface resources such as gas, petroleum and groundwater. The majority of the world's major river deltas are also reckoned to be subsiding because of the reduced sedimentation caused by upstream dams, and by compaction caused by the construction of infrastructure. These large local and regional variations mean that SLR in many areas of the world may be much larger than the forecast global mean.

In the western Pacific, recent studies have shown that several small low-lying islands have already disappeared. Australian researchers identified five vegetated reef islands in the Solomon Island group that have vanished since 1947, and a further six islands experiencing severe shoreline recession. Shoreline recession at two sites has destroyed villages that have existed since at least 1935 (Albert *et al.*, 2016).

Severe Storms

Severe storms generate surges of seawater across coastal areas and often cause extensive flooding inland. Their severity depends on the strength of the storm, its track, the depth of the ocean, and local hydrodynamic factors. Although there is uncertainty whether the frequency of storms is increasing globally, scientists agree that there has been an increase in the intensity of the strongest tropical cyclones in the North Atlantic over the last 40 years. There is a consensus that precipitation will increase together with maximum wind speeds. In other words, tropical cyclones are likely to get bigger and more powerful. Of the 51 SIDS included in this book, only four countries are usually considered to be completely safe from cyclones: Bahrain, Guyana, Suriname, and Guinea-Bissau. For others, for example Aruba, Trinidad and Tobago, and Singapore, the risk is low but not zero.

Winds and Waves

Changes in wave and wind patterns affect sedimentation and shoreline processes. The coastal impacts are also a function of wave direction and periodicity as well as the coastline itself. Long-period swells can pose a significant danger to coastal and offshore structures and shipping. Whether winds and waves will strengthen with climate change is uncertain and long-term trends are unclear.

Sea Surface Temperatures

Sea surface temperature (SST) has increased significantly during the past 30 years along more than 70% of the world's coastlines. However, there are large regional variations both geographically and seasonally. SST is increasing faster along coastlines than in the oceans themselves.

Freshwater Flows

Changes in river runoff arise from changes in precipitation, complex interactions between biophysical processes, as well as factors such as land-use change, water withdrawal rates, water retention infrastructure, and other engineered modifications to water courses.

A 2009 assessment of runoff trends for 925 of the world's largest rivers indicated that, over the period 1948 to 2004, only a third of the rivers showed statistically significant changes in flow. Of these, two-thirds showed a downward trend, and a third an upward trend (Dai *et al.*, 2008).

Although there is a degree of uncertainty around the forecasts, mean annual runoff is projected to increase in high latitude areas and in the wet tropics, and to decrease in most dry tropical regions. The changing patterns of freshwater flows and their unpredictability are primarily influenced by projected changes in precipitation, which is still a difficult factor to model at fine scale with any certainty.

Ocean Acidification

The chemistry of coastal marine systems is complex. Although the acidity of the oceans is projected to increase globally, there is considerable variation, especially in coastal waters. Seawater pH exhibits a much larger spatial and temporal variability in coastal waters compared with the open ocean due to the variable contribution of processes other than CO_2 uptake, such as upwelling intensity, deposition of atmospheric nitrogen and sulfur, the carbonate chemistry of riverine waters, as well as inputs of nitrogen and inorganic matter. Coastal acidification is forecast to continue to increase, but with large and uncertain regional and local variations.

Coastal Zones: Terrestrial and Intertidal Impacts

Beaches and Sand Dunes

Beaches and sand dunes, less common than rocky coasts, often exhibit distinct seasonal changes. They are highly valued for recreation, residences and tourism. Globally, beaches and dunes have in general undergone net erosion over the past century. Erosion is a complex process, and attributing erosion to climate change is difficult. Linking sea-level rise (SLR) to beach erosion is statistically significant in several places, but SLR may be only one of several factors at work and not necessarily the most important. Violent storms can erode and completely remove dunes, degrading land elevations and exposing them to inundation and further change if recovery does not occur before the next storm. Even in the absence of hard obstructions, barrier island erosion and narrowing can occur as a result of SLR and recurrent storms.

Coastal erosion is influenced by many factors: sea-level, currents, winds and waves (especially during storms, which add energy to these effects). Erosion of river deltas is also influenced by precipitation patterns inland, which change patterns of freshwater input, runoff, and sediment delivery from upstream. All the elements of coastal erosion are affected by climate change. A rise in mean sea-level generally causes the shoreline to recede inland due to coastal erosion. Increasing wave heights can cause coastal sand bars to move away from the shore and out to sea. High storm surges (sea-levels raised by storm winds and atmospheric pressure) also tend to move coastal sand offshore.

Higher waves and surges increase the probability that coastal sand barriers and dunes will be over-washed or breached. More energetic and/or frequent storms exacerbate all these effects. Changes in wave direction caused by climate change may produce movement of sand and sediment to different areas on the shoreline, changing the patterns of erosion.

Rocky Coasts

Rocky coasts with shore platforms form about three-quarters of the world's coasts and are characterized by strong environmental gradients, especially in the intertidal zone where climate change can potentially have large impacts. Cliffs and platforms have reduced resilience; once platforms are lowered or cliffs have retreated, it is difficult to rebuild them.

The range limits of many intertidal marine species have shifted by up to 50 km per decade over the last 30 years in the northern Pacific and northern Atlantic – much faster than most recorded shifts of terrestrial species.

Wetlands and Seagrass Beds

Vegetated coastal habitats and coastal wetlands (mangrove forests, salt marshes, seagrass meadows and macroalgae beds) extend from the intertidal to subtidal areas in coastal zones where they form key ecosystems.

There are complex synergies among coastal mangroves, coral reefs and seagrass beds: together, these habitats form diverse and structurally complex ecosystems in which the reef acts as a barrier that shelters seagrass beds and mangroves from high wave energy. In turn, seagrass and mangroves provide foraging and nursery habitats for many larvae and juveniles of reef species of fish and invertebrates, including those of commercial value to fisheries.

Vegetated coastal habitats are declining globally, rendering shorelines more vulnerable to erosion due to SLR and increased wave action, and leading to the loss of carbon stored in sediments. Recognition of the important consequences of the losses of these habitats for coastal protection and carbon burial has led to large-scale reforestation efforts in several countries.

Climate change is leading to range shifts in vegetated coastal habitats. Seagrass meadows are already under stress, particularly where maximum temperatures already approach their physiological limits. Heatwaves lead to widespread seagrass mortality. Kelp forests have declined in temperate areas in both hemispheres.

At the same time, ocean acidification is expected to enhance the production of seagrass, macroalgae, salt-marsh plants and mangrove trees through the fertilization effect of CO_2. Increased CO_2 concentrations may have already increased seagrass photosynthesis rates by 20%. However, there is only limited evidence suggesting that elevated CO_2 rates will increase seagrass survival rates or resistance to warming.

Coastal wetlands and seagrass meadows experience coastal squeeze in urbanized coastlines, with no opportunity to migrate inland with rising sea-levels. However, increased CO_2 and warming can stimulate marsh elevation levels, counterbalancing moderate increases in sea-level rise rates.

Climate change is projected to exacerbate the continued decline in the extent of seagrasses and kelps in the temperate zone, and the range of seagrasses, mangroves and kelp in the northern hemisphere will expand towards the pole. However, the limited positive impact of warming and increased CO_2 on vegetated ecosystems will be insufficient to compensate for the decline of their extent resulting from other human drivers such as land use change.

Coastal Aquifers

Coastal aquifers are important for the water supply of densely populated coastal areas, especially in small islands. Temperature and evaporation rise, precipitation changes and extended droughts affecting aquifer recharge contribute to saltwater intrusion on many islands. Sea-level rise and overwash from waves or storm surge have serious impacts on coastal aquifers, especially in low-lying areas. Excessive water extraction in coastal zones leads to salinization, and the increased usage of groundwater resources over the last century has led to a reduction in groundwater quality.

Estuaries and Lagoons

Coastal lagoons are shallow bodies of water separated from the sea by a natural barrier and connected at least intermittently to the ocean, while estuaries, where freshwater and seawater mix, are the primary conduit for nutrients, particulates and organisms from land to the sea.

Sediment accumulation in estuaries is directly affected by human drivers such as dredging and canalization, and indirectly by degraded watersheds, changes in sea-level, storminess, and freshwater and sediment supply from rivers. Coastal lagoons are also susceptible to alterations of sediment input and erosional processes driven by changes in sea-level, precipitation and storminess. Droughts, floods, and SLR impact upon estuarine circulation, tidal characteristics, suspended matter and turbidity, with consequences for coastal ecosystems and intertidal systems.

Coastal Marine Ecosystems

Coral Reefs

The economy and culture of island people is closely tied to their relationship with the oceans. Although that relationship has changed radically over the years, how island populations interact with their marine environment remains central to the future of island communities. That relationship is now threatened as sea-levels rise, coastal zones are eroded and inundated, and increasingly violent storms appear to be conveyed by the oceans themselves.

Coral reefs have always protected and provided sustenance to tropical islands. The reefs dissipate the energy of strong waves and storm surges; they attenuate coastal currents that would erode and damage tropical beaches. They are the essential habitat and breeding ground for a multitude of marine species that have provided food to island communities for millennia. Tropical islands and their coral reefs attract tens of thousands of visitors and bring prosperity to island communities that would struggle to develop without the lure of these natural features.

But these marine ecosystems are increasingly vulnerable, and the coral reefs around small islands are most at risk. The threats come from the island communities themselves: overfishing and destructive fishing, unregulated coastal development including ports and marinas, coastal engineering, waste-water runoff from hotels and tourism infrastructure, erosion upstream in poorly managed watersheds, nutrient fertilizer runoff from unsustainable agriculture, marine-based pollution from coastal shipping, and physical damage from anchors and boat groundings. All these activities damage and weaken coral reefs and reduce the ecosystem services they provide to coastal communities.

Climate change will worsen an already alarming situation. Coral cannot tolerate rising seawater temperatures: mass coral bleaching and mortality quickly follows positive temperature anomalies. The first widely recorded bleaching was in 1997–98, followed by bleaching events in 2002 and 2005. In 2015, another massive and widespread bleaching event started.

The increasing acidity of the oceans will reduce marine biodiversity and the rate of calcification of corals. The rate of dissolution of the reef structure will increase, and the carbonate balance of coral reefs will shift towards net dissolution. These changes will erode and damage the habitat for reef-based fisheries, increase the exposure of coastlines to waves and storms, and lessen the attraction of islands as destinations for international tourism.

A recent assessment of the threat to coral reefs conducted by the Washington-based World Resources Institute paints an alarming picture:

- More than 60% of the world's reefs are under immediate and direct threat from one or more local sources, such as overfishing, destructive fishing, coastal development, watershed-based pollution, and marine-based pollution.
- Overfishing and destructive fishing (using explosives or poisons) is the most pervasive immediate threat, affecting more than 55% of the world's reefs. Coastal development and pollution from watersheds each threaten about 25% of reefs. Marine pollution and damage from ships is widely dispersed, threatening about 10% of reefs.
- Approximately 75% of the world's coral reefs are rated as threatened when local threats are combined with thermal stress from rising surface water temperatures, linked to the widespread weakening and mortality of corals due to mass coral bleaching (Burke *et al.*, 2011, 2012).

The level of threat to the reefs in each of the SIDS regions is summarized below.

Pacific Region

The Pacific region is the largest of the three regions and holds more than a quarter of the world's coral reefs, most of which are found among the three major island groups of the western Pacific. In the northwest, the Micronesia islands consist of several archipelagos dominated by coral atolls but including several volcanic islands. Most of the reefs in this region are fringing reefs and barrier formations including New Caledonia's huge barrier reef, which is over 1300 km long. The Polynesian islands occupy an extensive area in the central Pacific including Tonga, French Polynesia, Samoa, Niue, Tuvalu, Kiribati and Nauru. Most of these islands are coral atolls interspersed with a few volcanic islands.

Papua New Guinea and the Solomon Islands comprise the eastern half of the Coral Triangle: a global biodiversity hotspot with more species of fish and corals than anywhere else on the planet. The islands and reefs to the east exhibit relatively low biodiversity, but provide habitat for large numbers of endemic species.

More than any other region on the planet, the people of the western Pacific live in close proximity to the coastal marine environment and the coral reefs adjoining and protecting their islands. The reefs are the foundation of local fisheries, while in many areas reefs provide important quantities of exported seafood. Tourism across the region is the mainstay of many island economies, particularly for small islands where alternative livelihoods are limited.

Although large areas of the Pacific still have relatively healthy reefs with good coral cover, the situation is changing. Overfishing is common and is linked to growing coastal settlements not only around the larger islands, but also in some of the smaller archipelagos in Micronesia. Watershed-based pollution on the higher-elevation islands affects a quarter of all reefs. In many areas, this is linked to clearing forests and erosion, but open-pit mining is also a significant source of sediments and pollution, most notably copper and gold mining in Papua New Guinea and nickel mines in New Caledonia. Coastal development affects almost a fifth of all reefs, most importantly in Fiji and Samoa.

In Papua New Guinea, sedimentation and pollution from inland areas are a threat to reefs. Natural phenomena have affected some islands, including outbreaks of crown of thorns starfish (COTS) in Papua New Guinea, the Cook Islands and French Polynesia. Coral bleaching events have been widespread – in Palau in 1998 and 2010, and around Kiribati in 2002 and 2005. In late 2010 this was caused by elevated sea surface temperatures across wide areas of Micronesia. In the eastern Pacific, the threats to reefs are more variable but they include the earliest recorded mass bleaching and coral mortality, which occurred in 1982–83.

Climate change impacts are projected to increase the proportion of threatened reefs up to 90% by 2030. Around Papua New Guinea, the Solomon Islands and Vanuatu, the combined effects of acidification with thermal stress will push many reefs into the very high or critical threat categories.

Caribbean Region

The Caribbean region includes about 10% of the world's coral reefs. Reef types include fringing and bank reefs as well as several barrier systems, most notably around Cuba and along the coast of Belize. The Bahamas group, including the Turks and Caicos Islands, is a large system of shallow banks with reefs at their outer edges. The most northern reefs are around Bermuda, warmed by the Gulf Stream.

The diversity of reef species is relatively low. While there are more than 750 species of reef building corals across the Indian and Pacific Oceans, the Caribbean hosts less than 65.

However, these species tend to be unique, with over 90% of fish, corals and crustaceans found nowhere else.

The Caribbean region is densely populated: the SIDS islands support a total population of over 40 million. Although many islands are relatively wealthy, there remains a heavy dependence on the reefs for food and tourism. In many of the poorer islands, tourism has long surpassed the contributions of agriculture and industry to GDP. Even in places where tourism is less intense, the reefs play an important supporting role: providing food, protecting coastlines, and providing sand for beaches.

The region is exposed to regular and intense tropical storms, and numerous coastal settlements are protected by barriers of coral reefs, which break waves far offshore and reduce the effects of coastal flooding and erosion.

The reefs in the Caribbean have been in decline for several decades. Since the 1980s, a major cause of damaged reefs has been the impact of diseases, particularly those affecting long-spined sea urchins and many coral species. Urchins are important herbivores, feeding on reef algae that would otherwise smother coral. This role is even more important in areas where overfishing has depleted herbivorous fish. Disease has also killed off extensive areas of elk-horn and stag-horn corals – once widespread in the Caribbean and important reef-building corals. The causes are poorly understood, but anthropogenic stresses such as pollution are thought to be an important factor. Overfishing is endemic, and the reefs have some of the lowest biomass values anywhere in the world. Groupers and snappers are rare, and fishers have moved down the food chain, taking increasing numbers of the herbivores that keep coral healthy.

Reefs have survived overfishing, but the combination of overfishing, pollution and bleaching has been devastating for the reefs of Jamaica, Haiti, and many of the islands in the Lesser Antilles. The 2005 bleaching event in the US Virgin Islands was followed by outbreaks of disease that led to further losses of coral. Tropical storms and hurricanes also damage reefs. Twelve hurricanes and eight tropical storms hit the northern Caribbean between 2004 and 2007. The loss of living coral in the Caribbean has led to a weakening of the structure of many reefs, further reducing biodiversity and productivity.

More than 75% of the reefs in the Caribbean are threatened, with more than half in the high and very high threat categories. Overfishing is the predominant threat, followed by coastal development and watershed-based pollution – the latter threatening at least a quarter of the reefs. Only the more remote reefs are better protected: those around the smaller islands of the Bahamas and the coastal systems of the southwestern Caribbean.

Indian Ocean Region

The SIDS in this region are the Comoros islands, the Seychelles, Mauritius and the Maldives. In the Maldives, the islands are built from reefs and the people depend heavily on fishing and tourism. Reef-based tourism is a mainstay of the island economies.

The devastating bleaching event of 1998 hit this region harder than any other. In the Maldives and the Seychelles more than 80% of coral suffered complete mortality. Further bleaching occurred in 2001 and 2005, although many reefs recovered. More than 65% of reefs in the Indian Ocean are at risk from local threats, with a third rated at high or very high risk. The risk is high along the mainland shores. The single biggest threat is overfishing, which affects at least 60% of coral reefs. Dynamite fishing is also a problem, occurring mainly along the coast of Tanzania. The threat in the Maldives has intensified due to overfishing, probably linked to the increase in population. Although fishing in the Maldives targets deep-water species such as tuna, they still depend on bait fish caught on the reefs.

Mangroves

Mangrove forests, made up of salt-tolerant trees and shrubs, play a vital role in erosion and flood control, fisheries support, carbon storage, biodiversity conservation and nutrient cycling. Many coastal communities rely on mangroves for food, forest products and tourism revenue, and the forests provide a natural coastline defence by reducing wave and wind velocity.

Mangrove forests are everywhere in retreat. Asia is the epicentre of global mangrove loss, with annual loss rates nearly double the global average. More than half of the mangrove loss over the past decade occurred in the Asia Insular region, comprising Indonesia, Malaysia, Papua New Guinea, Solomon Islands, Brunei, Timor-Leste, Singapore and the Philippines. This region also boasts the world's largest mangrove area, with high biodiversity and enormous carbon storage potential. It is particularly suitable for mangrove growth because of its extensive coastlines, numerous islands, high rainfall and significant freshwater sources.

The aquaculture industry is one of the fastest-growing animal-producing sectors in the world, and as it has grown, so have concerns about its impact on mangroves and other ecosystems. In the 1980s and 1990s, a largely unregulated boom in shrimp aquaculture led to a clearing of a significant area of mangroves for aquaculture ponds.

Agricultural expansion is also a major cause of mangrove loss in Asia, mainly through the encroachment of palm oil plantations into mangrove forests. Oil spills and other chemical pollution, such as fertilizers and pesticides from nearby farms, also pose threats to Asian mangroves, while growing populations increasingly degrade mangrove habitats through coastal and urban development.

Conservation efforts for mangrove forests on many islands are minimal, and only a few mangrove forests are incorporated within legally protected areas.

As sea-levels rise, mangrove forests move further inland, maintaining the same height relationship with mean sea-level and tides. Where the coastal zones are built up, or where hard engineered structures have been installed to protect against erosion, storm surges and flooding, mangroves cannot populate more elevated areas and will die off. Climate change can therefore result in a reduction in coastal mangrove forests unless there is space into which the mangroves can move. If mangrove forests die, coastal fisheries will suffer even further deterioration.

Recent research has determined that mangrove forests store huge amounts of carbon: much more than tropical forest and other coastal ecosystems. Figure 2.2 shows that mangroves store significantly more carbon per hectare than other ecosystems. Many countries are now including the reforestation of mangroves as a way to increase carbon sequestration while at the same time reinforcing coastal protection (USAID, 2016).

Biodiversity

Islands are home to an incredible wealth of biodiversity. The isolation of islands over millennia has led to the evolution of unique species that are found nowhere else on Earth.

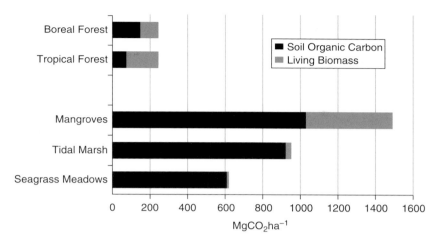

Figure 2.2 Mangroves store huge amounts of carbon. *Source:* USAID Climate Action Review 2010–2016.

Islands hold higher concentrations of endemic species than continents, and the number and proportion of endemic species increases with the isolation of the island, its size, and the variety of its habitats. Over 90% of Hawaiian island species are endemic. In Mauritius, about half of all higher plants, mammals, birds, reptiles and amphibians are endemic, and the Seychelles has the highest level of amphibian endemism in the world. Cuba is home to 18 endemic mammals, while mainland Guatemala and Honduras have only three each. Madagascar, the fourth largest island in the world, is home to more than 8000 endemic species.

The unique characteristics that make island biodiversity so exceptional also make it particularly fragile and vulnerable. Island species often develop survival strategies based on interdependency, co-evolution, and mutualism, rather than defence mechanisms against predators as competitors. Thus, many island species have become rare or threatened, and islands have a disproportionate number of recorded species extinctions compared with continental ecosystems. Of the over 700 recorded animal extinctions that have occurred in the last 400 years, about half were of island species. At least 90% of the bird species that became extinct during this period were island species.

In the Caribbean, where the distance between islands is relatively small compared with the Pacific, groups of islands link biodiversity corridors especially for birds. A biodiversity 'hotspot', the region supports a wealth of biodiversity within its diverse terrestrial ecosystems, with a high proportion of endemicity, making the region biologically unique. It includes about 11,000 plant species, of which 72% are endemics. For vertebrates, high proportions of endemic species characterize the herpetofauna, probably due to low dispersal rates, in contrast to the more mobile birds and mammals. Much less diversity in marine species exists because of the high degree of connectivity. The Caribbean Current runs through the basin all year long, transporting larvae between the islands. As a result, marine habitats share many of the same species, in contrast to the region's terrestrial biodiversity with its high rates of endemism (CEPF, 2010).

However, the Caribbean's biodiversity is at serious risk of species extinction. More than 700 species are threatened, making the Caribbean one of the top hotspots assessed by the CEPF for globally threatened species. The region has by far the highest percentage of threatened or extinct amphibian species. In fact, the top five countries in the world with the highest levels of threatened and extinct amphibians are all in the Caribbean.

Biodiversity is crucial to food security in many small isolated islands and especially among the SIDS. The continental shelves and coastal ecosystems of many SIDS are of major economic importance for settlements, subsistence and commercial agriculture, fisheries and tourism.

Worst-case Scenario: The Reefs of Haiti

In 1926, the New York Zoological Society funded a major marine biological expedition to Haiti and filmed the coral reefs swarming with large fish (Beebe, 1928). However, extensive overfishing in Haiti has been prevalent since at least the 1980s, and the situation has become progressively worse (Miller *et al.*, 2007). Reef fishermen initially target desirable predators such as grouper and snapper. As these species are fished out, fishermen turn to herbivores such as parrotfish and doctorfish – with disastrous consequences for the health of the reefs.

Between 2011 and 2017, Reef Check carried out over 300 surveys of reefs covering 90% of the 1700 km long coast of Haiti. During these underwater surveys of Haiti's coast, fish were rarely counted with a total length greater than 20 cm. In shallow waters less than 15 m deep, only four fish larger than 50 cm were observed during surveys of almost the entire coast of Haiti. Reef Check scientists concluded that the coral reefs in Haiti are the most overfished they have seen in the 90 coral reef countries where they work throughout the world. The overfishing of herbivores has destabilized most of Haiti's reefs, and they have transitioned from coral-dominated to algae- and sponge-dominated reefs (Reef Check, 2016).

The coastal areas surveyed in Haiti are shown in Figure 2.3 as a green line.

The results of the Reef Check surveys in Haiti show that the average living coral cover on Haiti's coral reefs was 15% as measured by standard line-intercept surveys on reefs, and less than 10% when measured by wide area surveys using the manta tow that included back-reef and non-reef areas (e.g. lagoons). This is about half the mean value for the Caribbean, which is around 25%. The cover of macroalgae on reefs in Haiti is 30%, a level more than double the Caribbean mean (13%). The cover of sponge in Haiti is about 10%, a value above the Caribbean mean of 7%. These results confirm that the reefs of Haiti have been destabilized due to nutrient inputs that fertilize algal growth, and overfishing that has reduced the population of herbivorous fish, as well as the die-off of the herbivorous *Diadema* sea urchin some 30 years ago.

The survey results confirm that very low numbers and extremely small sizes of all indicator fish were recorded during surveys in all areas of the country. For analysis, individual surveys from each area of the country were grouped together into 15 areas. Out of the 15 areas (groups of sites), less than one fish was counted per 100 m^2 of reef at seven sites. There are almost no grouper, and the Nassau grouper was only seen in Fort Liberté.

While overfishing is a global problem, these fish abundance results are the lowest recorded by Reef Check anywhere in the world. Based on size, all indicator fish counted were juveniles less than 20 cm fork length except for moray eels. Some large schools of

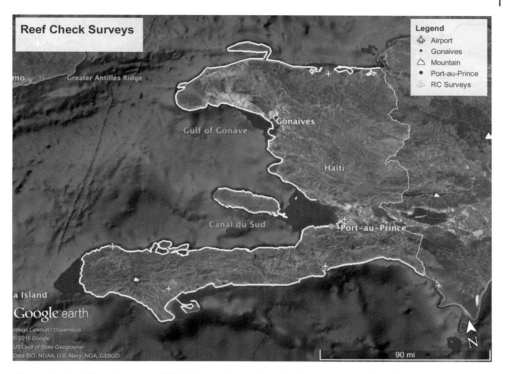

Figure 2.3 Reef Check surveys of Haiti. *Source:* Reproduced with permission of Reef Check International. (*See insert for color representation of the figure.*)

mature Haemulids (grunts) and Sparids were observed below 20 m depth at St Marc Point. Only four barracuda and one eagle ray were observed during manta tows of the entire country.

Butterflyfish have been collected for the aquarium trade for many years in Haiti, and are also eaten when caught in gills nets and traps. The snapper, grouper, grunts and moray eel are all predators, while parrotfish are a key herbivore. On a healthy Caribbean reef, a Reef Check survey would normally count about 10 to 20 mature grouper, dozens of parrotfish, snapper, grunts, butterflyfish, and a few moray eels. No mature pelagic fish such as jacks, dorado or tuna were observed on any survey.

The results from the Reef Check surveys indicate that 'indicator invertebrates' are also extremely low in number – near zero for most of them. These results mirror those for fish and indicate overfishing of most of these organisms. The banded coral shrimp is collected for the aquarium trade, while the pencil urchin, triton, and flamingo tongue are collected and sold for the curio trade. The collector urchin and spiny lobster are collected for food. The long-spined black *Diadema* sea urchin is an important ecological indicator. The Haitian Government banned the collecting of triton shells many years ago, but there is no enforcement of this ban.

Based on the quantitative Reef Check survey data, it has been shown that the reefs of Haiti are the most overfished and destabilized in the world. Many reefs in Haiti resemble ghost towns, with algal-dominated reefs and only a few tiny fish (see Figure 2.4).

Figure 2.4 A typical former coral reef at La Gonave in Haiti overfished to the point where it has become an algal dominated reef with a few stubs of coral surviving but no fish. *Source:* Reproduced with permission of Reef Check International. (*See insert for color representation of the figure.*)

Haiti is an extreme example of the impact of overfishing on the marine environment. There are many reasons why Haiti's environment is in such poor shape. Political instability, widespread corruption, high population densities, mountainous terrain, intractable poverty, and plain bad luck have all played a part. But like many islands where agriculture cannot provide an adequate livelihood for the majority of the population, coastal communities have looked to the sea to provide food and sustenance. But there is a limit to the amount of fish and seafood that the coastal environment can provide, and once this limit is reached and exceeded, the marine environment is degraded and eventually destroyed. Climate change will make it even harder for coral reefs and the marine ecosystems to recover; higher sea surface temperatures, greater seawater acidity, and extreme weather will further stress marine ecosystems unless they are closely managed and carefully protected – a policy that Haiti has never managed to implement successfully.

Low Elevation Coastal Zones

If the oceans remained calm and tranquil, sea-level rise, being generally in the range of 2–3 mm per year, would not pose much of an immediate threat to small islands. There would be time to take the necessary measures to protect coastal communities and infrastructure. However, the oceans are in constant motion, and when powerful storms drive waves to heights of several metres above normal levels, they can cause huge amounts of damage.

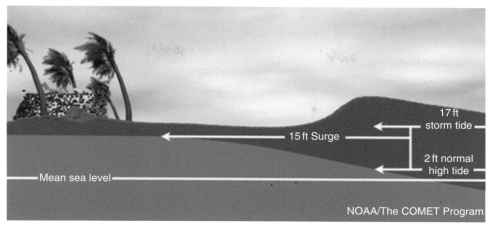

Storm Surge vs. Storm Tide

Figure 2.5 Storm surge and storm tide. *Source:* National Hurricane Center.

Storm surge is an abnormal rise of water generated by a storm, over and above the normal height of the regular tides. The term 'storm tide' is defined as the seawater-level rise due to the combination of storm surge and normal tide, as depicted in Figure 2.5.

This rise in seawater level can cause extreme flooding in coastal areas. Storm surges can easily reach 8 metres. Table 2.5 shows some of the more destructive storm surges recorded in the US during the last 25 years (National Hurricane Center, 2015a)

These storm surge levels were recorded in the US. Reports on tropical cyclones rarely mention the height of the storm surge, but the extensive coastal and inland flooding caused by many of these cyclones is testimony to the high storm surges often generated by these extreme events. Hurricane Matthew, which pummelled Jamaica, Haiti, Cuba and the Bahamas in October 2016, generated a storm surge of between 3 to 5 metres approaching the Bahamas (National Hurricane Center, 2016).

In January 2016, a hurricane named Alex formed in the northern Atlantic – an unusual event for that time of the year. Although not an especially fierce storm, it reportedly produced a storm surge of 18 metres. A red alert was issued for five of the Azores' nine islands. It was noted at the time that seawater temperatures were about 2 °C higher than normal.

Recent advances in satellite imagery have enabled scientists to estimate more accurately the number of people and the extent of the coastal area at different elevations above mean sea-level. People that live only a few metres above mean sea-level in areas of the world where cyclones and hurricanes are frequent obviously face an extremely high risk of being hurt or killed by storm surges associated with these extreme events.

The small island states where populations are most at risk are shown in Table 2.6. The population threshold in this table is arbitrarily set at 50,000 to highlight the islands where the greatest number of people in low elevation coastal zones (LECZs) are most at risk (CIESIN, 2013).

These are huge numbers. Singapore has more than a third of a million people living within 1 metre of mean sea-level. Due to the diminishing Coriolis force towards the equator, the equatorial zone is generally cyclone-free. However, on 27 December 2001,

Table 2.5 Storm surges recorded in the US this century (National Hurricane Center, 2015b).

Storm and year	Details	Storm surge height (m)
Sandy 2012	Second only to Hurricane Katrina in terms of the damage wrought in the US, in Jamaica the hurricane accounted for J$9.7 billion or 0.8% of 2011 GDP in direct and indirect damage. In Cuba there was extensive coastal flooding and wind damage inland, destroying some 15,000 homes, killing 11, and causing $2 billion (2012 USD) in damage. Sandy also caused two deaths and damage estimated at $700 million (2012 USD) in the Bahamas.	4–5
Ike 2008	The Category 2 hurricane made landfall near Galveston, Texas, leaving a trail of death and destruction. It is estimated that flooding and mudslides killed 74 people in Haiti and two in the Dominican Republic, compounding the problems caused by the previous storms Fay, Gustav and Hanna. The Turks and Caicos Islands and the southeastern Bahamas sustained widespread damage to property. Seven deaths were reported in Cuba.	6–7
Rita 2005	Rita developed in September and swept across the Florida Keys as a category 2 hurricane. Moving westwards, it made landfall again on the east Texas coast. Devastating storm surge flooding occurred in SW Louisiana and SE Texas.	3–5
Katrina 2005	A tropical depression formed on August 23 about 200 miles southeast of Nassau in the Bahamas. Moving northwestward, it became Tropical Storm Katrina during the following day about 75 miles east-southeast of Nassau. The storm moved through the northwestern Bahamas on August 24–25, and then turned westward toward southern Florida. Katrina made landfall in the New Orleans area of the Mississippi coast and caused massive and extensive flooding and many deaths.	8–9
Dennis 2005	Dennis caused 42 deaths: 22 in Haiti, 16 in Cuba, 3 in the US, and 1 in Jamaica. The hurricane caused considerable damage across central and eastern Cuba as well as the western Florida Panhandle, including widespread utility and communications outages. Extensive storm surge-related damage also occurred near St Marks, Florida, well to the east of the landfall location. The damage associated with Dennis in the US is estimated at $2.23 billion.	2–3
Ivan 2004	Ivan developed off the coast of Africa on August 31. Ivan strengthened to Category 5 force south of the Dominican Republic, passing close to Jamaica before making landfall in the US in Alabama. Ivan completely over-washed the island of Grand Cayman where 95% of buildings were destroyed. The central mangrove area was swamped with seawater which eventually poisoned vast areas of virgin mangrove swamp (CEPF, 2010). In Jamaica, Hurricane Ivan caused losses equivalent to 8.0% of GDP.	3–5
Frances 2004	Originating southwest of Cabo Verde islands on August 25, Frances produced a storm surge of about 2 metres at its Florida east coast landfall. Heavy rains and the resulting flooding caused damage estimated at almost $9 billion.	2
Charley 2004	Charley developed southeast of Barbados on August 9, and became a hurricane on August 11 near Jamaica. The hurricane came ashore on the west coast of Florida as a Category 4 storm. Charley was a small hurricane but still did substantial damage in Florida, N. Carolina and Virginia.	2–3

Table 2.5 (Continued)

Storm and year	Details	Storm surge height (m)
Isabel 2003	A tropical wave that exited the African coastline on September 1 developed into Tropical Storm Isabel on the morning of September 6. Isabel became a hurricane on September 7 and rapidly intensified to Category 4 hurricane strength while the eye was located more than 1100 miles to the east of the Leeward Islands. This impressive hurricane reached Category 5 strength on September 11, making Isabel the strongest hurricane in the Atlantic basin since Mitch in October 1998.	2–3

Table 2.6 Coastal populations (thousands) at different elevations above mean sea-level (metres).

SIDS	1 m	3 m	5 m	10 m
Singapore	344.7	440.1	551.5	1590.9
Guinea Bissau	138.1	148.1	254.6	560.8
Maldives	63.5	89.9	162.9	307.9
Puerto Rico	51.7	195.8	328.8	550.4
Guyana		132.8	221.8	279.8
Cuba		124.0	326.9	1014.4
Suriname		108.6	300.7	421.0
Haiti		71.2	123.3	590.2
Dominican Republic		51.4	123.3	575.5
Jamaica			91.4	221.5
Bahamas			75.5	308.0
Belize			66.7	117.3
Papua New Guinea				164.8
Trinidad and Tobago				155.1
Fiji				107.6
New Caledonia				71.6
Mauritius				68.2
Kiribati				65.9
US Virgin Islands				52.0
Total LECZ populations	**598**	**1362**	**2627**	**7223**

LECZ, low elevation coastal zone.

tropical cyclone Vamei formed near latitude 1.4°N, and made landfall about 60 km northeast of Singapore. Although this event did not cause any major destruction to Singapore, the neighbouring states of Johor and Pahang in Malaysia suffered inland flooding and landslides caused by the heavy precipitation of the storm. Vamei was unusual in that it formed so close to the equator. But climate change is affecting weather patterns worldwide, and clearly Singapore is at extreme risk if tropical cyclones at low latitudes become more frequent. Even without the advent of tropical storms, Singapore is vulnerable; sea-level rise alone will eventually inundate coastal areas where hundreds of thousands of people live.

Guinea-Bissau is not in a region where tropical cyclones occur. However, all the other countries on the list above are frequently impacted by hurricanes in the Caribbean or typhoons in the Pacific. Cuba and Puerto Rico look particularly exposed. Both countries have almost a third of a million people living in low-lying areas less than 5 metres above mean sea-level.

In terms of the percentage of populations at risk, the picture is different, since the islands with smaller populations do not show up in the table above. Table 2.7 shows islands where more than 20% of the population is within the indicated LECZ.

What is striking about these data is the large percentage of the populations within 5 metres of mean sea-level, and therefore well within the reach of storm surges associated with hurricanes and typhoons. In Tuvalu and the Maldives, almost half the population is living less than 5 metres above mean sea-level. In Suriname, a relatively large country, over half the population is living on the coast. If an island is large enough, people can be quickly moved inland and out of harm's way, but infrastructure cannot be moved so easily.

If the data are arranged to show the area of land that is likely to be flooded by storm tides, the picture looks a little different (see Table 2.8). A significant fraction of these low elevation coastal zones is likely to be urban areas which, if flooded, will

Table 2.7 Coastal populations (percentages) at different elevations above mean sea-level (metres).

SIDS	1 m	3 m	5 m	10 m
Tuvalu	20.3	34.1	47.5	74.5
Maldives		26.6	48.2	91.1
Marshall Islands		20.7	35.9	80.7
Suriname		20.3	56.2	78.7
Guyana			27.9	35.2
Belize			20.6	36.2
Kiribati			20.5	65.2
Bahamas			20.3	82.8
Seychelles				45.8
Guinea Bissau				33.7
Singapore				30.0
N. Mariana Islands				29.5
New Caledonia				28.3
Anguilla				24.6
Antigua & Barbuda				21.8
British Virgin Islands				20.0

Table 2.8 Islands that have more than 10% of land area at risk at LECZ3 (low elevation coastal zone – 3 metres).

SIDS	Total land area, km^2	Land area at LECZ3, km^2	Percentage of land area at risk
Anguilla	91	10	12.8
Bahamas	13,940	3353	24.0
British Virgin Islands	151	29	19.2
Cabo Verde	4033	95	23.6
French Polynesia	4000	512	12.8
Kiribati	726	249	34.3
Maldives	300	86	28.7
Marshall Islands	181	57	31.5
Seychelles	457	52	11.4
Tuvalu	26	6	23.1

result in substantial loss of economically essential and productive infrastructure, as well as the probable shutdown of local government services and communication systems.

All land on islands within 3 metres of present mean sea-levels is highly vulnerable to catastrophic flooding from storm surges. Six island states – the Bahamas, Cabo Verde, Kiribati, Maldives, Marshall Islands and Tuvalu – have more than 20% of their land area exposed to this threat. Kiribati once again looks hugely exposed. It is certain that the majority of tourism infrastructure is constructed on these coastal lands and, if flooded, the economic impact of this loss of infrastructure will be catastrophic. For the Bahamas and the Maldives, tourism receipts as a percentage of GDP are estimated at 28% and 77% respectively. The exposure to sea-level rise and extreme weather for the Maldives is well captured by the photograph (Figure 2.6) of the capital Malé on North Malé atoll in the Maldives.

The value of the coastal infrastructure at risk is of course substantial, as many islands have invested considerable amounts of money in beach hotels and associated infrastructure to support tourism. A study in 2015 of 12 Pacific island states (Kumar & Taylor, 2015) found that 57% of the infrastructure identified in the study was within 500 metres of the coast. The total replacement value of all the infrastructure was estimated at $27.7 billion, 79% of which was within the 500-metre coastal strip.

For Kiribati, 97% of all infrastructure falls within 500 m of the coast, and 67% within 100 m. For the Marshall Islands, 98% of infrastructure is within 500 m and 72% within 100 m. Similar numbers are found in Tuvalu. In terms of replacement value, the data are even more striking: Kiribati, Marshall Islands and Tuvalu have 95%, 98% and 99%, respectively, of built infrastructure by value within 500 m of the coast, indicating that almost all built infrastructure by value is located along the coast. For Vanuatu, although 52% of built infrastructure is beyond the 500 m band, this represents only 10% by value, indicating that the most valuable infrastructure, which includes ports and in some cases oil refineries, is on the coast.

Figure 2.6 The island of Malé in the Maldives. *Source:* https://en.wikipedia.org/wiki/Mal%C3%A9#/media/File:Male-total.jpg. Used under CC BY-SA 3.0. (*See insert for color representation of the figure.*)

Although sea level rise and storm surges are major threats, ocean swell caused by weather events thousands of kilometres away can cause substantial damage to coastal communities in the tropics. In 1987, long-period swells originating in the Southern Ocean 6000 km from the Maldives caused flooding, property damage, destruction of sea defence structures, and serious coastal erosion (Nurse *et al.*, 2014). In 2008, swells generated in the north Pacific Ocean overwashed lowlying islands in the Pacific, causing severe damage to housing and infrastructure that affected about 100,000 people across the region (Hoeke *et al.*, 2013).

Climate-related disasters

Climate-related disasters are on the rise worldwide, causing loss of life and destruction, and setting back economic and social development by years at a time. From 1970 to 2012, 8835 disasters, 1.94 million deaths, and US$2.4 trillion of economic losses were reported globally as a result of hazards such as droughts, extreme temperatures, floods, tropical cyclones and related health epidemics.

Storms and floods accounted for 79% of the total number of disasters due to weather, climate and water extremes, and caused 55% of lives lost and 86% of economic losses between 1970 and 2012. Droughts caused 35% of lives lost, mainly due to the severe African droughts of 1975 and 1983–84.

The 1983 drought in Ethiopia ranked top of the list in terms of casualties, claiming 300,000 lives, about the same number as Cyclone Bhola in Bangladesh in 1970. Drought in Sudan in 1984 killed 150,000 people, whilst the cyclone locally known as Gorky killed almost 140,000 people in Bangladesh in 1991.

Hurricane Katrina in the US in 2005 caused the worst economic losses, at US$146.89 billion, followed by Sandy in 2012 with a cost of $50 billion.

The worst ten reported disasters in terms of lives lost occurred primarily in least developed and developing countries, whereas the economic losses were mainly in more developed countries. The socioeconomic impact of disasters is escalating because of their increasing frequency and severity and the growing vulnerability of human societies.

Africa: From 1970 to 2012, there were 1319 reported disasters causing the loss of 698,380 lives and economic damage of US$26.6 billion. Although floods were the most prevalent type of disaster, droughts led to the highest number of deaths. The severe droughts in Ethiopia in 1975 and in Mozambique and Sudan in 1983–84 caused the majority of deaths. Storms and floods, however, caused the highest economic losses.

Asia: Over 2600 disasters were reported in the 1970–2012 period, causing the loss of 915,389 lives and economic damage of US$789.8 billion. Most of these disasters were attributed to floods and storms. Storms had the highest impact on life, causing three-quarters of the lives lost, while floods caused the greatest economic loss. Three tropical cyclones were the most significant events, striking Bangladesh and Myanmar and leading to over 500,000 deaths. The largest economic losses were caused primarily by disasters in China, most notably by the 1998 floods.

South America: From 1970 to 2012, South America experienced 696 reported disasters that resulted in 54,995 lives lost and US$71.8 billion in economic damages. With regard to impacts, floods caused the greatest loss of life (80%) and the most economic loss (64%). The most significant event during the period was a flood and landslide/mudslide that occurred in Venezuela in late 1999, and caused 30,000 deaths.

North America, Central America and the Caribbean reported 1631 disasters that caused the loss of 71,246 lives and economic damages of US$1008.5 billion. The majority of the reported disasters in this region were attributed to storms and floods. Storms were reported to be the greatest cause of lives lost and of economic loss.

The South-West Pacific region experienced 1156 reported disasters in 1970–2012 that resulted in 54,684 deaths and US$118.4 billion in economic losses. The majority were caused by storms and floods. The most significant reported disasters with regard to lives lost were tropical cyclones, mainly in the Philippines, including the event of 1991, which killed 5956 people. The 1981 drought in Australia caused US$15.2 billion in economic losses, and the 1997 wildfires in Indonesia caused nearly US$11.4 billion in losses.

In Europe, 1352 reported disasters claimed 149,959 lives and caused US$375.7 billion in economic damages. Floods and storms were the most reported cause of disasters, but extreme temperatures led to the highest proportion of deaths, with 72,210 lives lost during the 2003 western European heatwave and 55,736 during the 2010 heatwave in the Russian Federation. In contrast, floods and storms accounted for most of the economic losses during the period.

Source: Adapted from WMO (2014).

Natural disaster management

SIDS are located in zones that are highly susceptible to and affected by natural disasters, including hurricanes, droughts and floods. Over the period 1990–2013, SIDS overall were affected by 491 natural disasters, predominantly storms and floods, but including epidemics, earthquakes, droughts and volcanic eruptions.

The impacts of these disasters can be devastating both in human terms and in terms of the cost to national economies. Between 1990 and 2013 the highest values recorded for SIDS was around $8 billion for the estimated damages in 2010. Hurricane Sandy in 2012 killed 11 people in Cuba and caused $2 billion in damages. Jamaica and the Bahamas also suffered significant damage from Sandy. The Pacific region may be even more vulnerable and has been dubbed the most disaster-affected region in the world. In December 2012, tropical cyclone Evan battered Fiji and Samoa, displacing more than 5000 people, stranding tourists, and costing millions of dollars in damage. It was the worst cyclone recorded in the area in over 20 years, and the island is still rebuilding from the damage.

The population of the SIDS affected directly by natural disasters in 2001–10 rose significantly compared with the previous decade. This increase is thought to have been driven by both a stronger migration towards coastal and urban areas, and the increasing frequency of extreme weather events. Between 2001–10, the total estimated cost of damage was $22 billion – almost triple the figure of the previous decade.

Source: UNDESA (2014).

Agriculture

The majority of the small island developing states are heavily dependent on agriculture – where the agricultural population is defined as all people that depend on agriculture, forestry, fishing and hunting for their livelihoods (FAO, 2014).

Table 2.9 The importance of agriculture in the SIDS.

Small island developing state	Agricultural population, share of total		Agricultural value added by share of GDP 2009–2011, %
	2000, %	2010, %	
Comoros	73.6	69.4	46.3
Guinea-Bissau	82.5	79.3	–
Haiti	64.1	58.8	–
Papua New Guinea	78.2	72.7	35.9
Sao Tome & Principe	92.3	88.8	–
Seychelles	76.4	74.0	1.8
Solomon Islands	71.7	67.6	38.9
Timor-Leste	81.2	78.3	–

In 30 of the SIDS, at least a fifth of the population is engaged in agriculture. Even where agriculture's contribution to GDP is small, a substantial part of the population may still be involved. For instance, in the Bahamas and Antigua-Barbuda, agriculture makes about the same contribution to GDP, but as a percentage, Antigua-Barbuda has nine times as many people working in the agricultural sector as the Bahamas. In the Seychelles, agriculture makes a contribution of less than 2% to GDP, yet almost three-quarters of the population is engaged in agriculture.

As Table 2.9 shows, in eight of the SIDS, the majority of the population are engaged in agriculture: Comoros, Guinea-Bissau, Haiti, Papua New Guinea, Sao Tome and Principe, Seychelles, Solomon Islands and Timor-Leste. It is also noticeable that in every country without exception, the fraction of the population engaged in agriculture has decreased significantly over the period 2000–2010. However, in absolute terms, due to the growing population on each island, the number of people engaged in agriculture may well be rising (FAO, 2013).

Globally, more than 70% of agriculture is rain-fed. Agriculture is therefore very sensitive to changes in rainfall, and highly vulnerable to climate change impacts that will change patterns of precipitation and increase local temperatures, and to weather that becomes become increasingly variable and frequently extreme. However, the resilience of agriculture in the SIDS can be increased using conservation agriculture techniques and 'climate-smart' practice. This is discussed further in Chapter 3.

Fisheries

Millions of coastal people across the globe depend on fishing both for food and as a source of income. Often combined with other ways to make a little money (agriculture and commerce being foremost), the revenue from fishing is essential for many poorer families and particularly for those living on small islands where nearly all coastal communities regularly engage in fishing to some degree. The fishing, aquaculture, and fish production sector employs around 150 million people in developing countries; of this total, 38 million are fishing full-time. The number of full-time fishers has been growing steadily since 1990, at a rate ten times higher than the increase in agriculture workers over the same period (World Bank, 2005). This growth in employment has been mainly in small-scale fisheries in developing countries. Many coastal fishers are among the poorest sector of society. Fishing is a way of reducing the vulnerability of poor families by supplementing and diversifying their incomes. Fishing is often the last-resort livelihood for the poor.

Climate change will seriously impact fisheries on small islands. It will change both the productivity of fishing areas and the distribution of fish in marine and inland waters. Table 2.10 outlines the types of changes that can be expected and potential impact on fisheries (FAO, 2007).

What is clear from the analysis outlined in this table is that both the productivity of small island fisheries and the distribution of fish resources in both marine and inland waters is going to change significantly. The impact on small island developing states is likely to be severe.

Nutrition will almost certainly suffer. The importance of fish in terms of nutrition is particularly high in developing countries, where protein intake may be insufficient.

Table 2.10 Climate change impacts on fisheries.

Type of change	Climatic variable	Impact	Potential outcome for fisheries
Physical environment	Changes in acidity caused by higher levels of CO_2	Effects on calciferous animals such as mollusks, crustaceans, corals and echinoderms	Potential declines in production for calciferous marine resources
	Warming of the ocean's upper layers	Warm-water species replacing cold-water species Planktonic species move to higher latitudes	Shifts in distribution of plankton, invertebrates, fishes and birds towards the poles
		Timing of phytoplankton blooms shifts and zooplankton composition changes	Potential mismatch between prey (plankton) and predator (fish) leads to decline in production and biodiversity
		Coral bleaching and reefs dying	Reduced production of coastal fisheries
	Sea-level rise	Loss of coastal fish breeding and nursery habitats, e.g. mangroves	Reduced production of coastal and related fisheries
Fish stocks	Higher water temperatures, changes in ocean currents	• changes in sex ratios • altered timing of spawning • altered timing of migration • altered time of peak abundance • species poleward shifts	Possible impacts on timing and levels of productivity across marine and freshwater systems
		Increased invasive species, diseases and algal blooms	Reduced production of target species in marine and freshwater systems
		Changes in fish recruitment success	Abundance of juvenile fish affected and therefore production in marine and freshwater
Ecosystems	Reduced water flows and increased droughts	Changes in lake water levels Changes in dry water flows in rivers	Reduced lake productivity Reduced river productivity
	Increased frequency of ENSO events	Changes in timing and latitude of upwelling Coral bleaching and die-off	Changes in pelagic fisheries distribution Reduced productivity in coral reef fisheries
Coastal infrastructure and fishing operations	Sea-level rise	Coastal profile changes, loss of harbours, homes, and tourism investments Increased exposure of coastal areas to storm events	• Costs of adaptation make fishing less profitable • Risk of storm damage increases costs of insurance and/or rebuilding • Coastal household's vulnerability increases

Table 2.10 (Continued)

Type of change	Climatic variable	Impact	Potential outcome for fisheries
	Increased frequency and intensity of storm	More days at sea lost due to bad weather and risk of accidents increases	• Increased risks of both fishing and coastal fish-farming, making them less viable livelihood options for the poor
		Aquaculture installations on the coast more likely to be damaged or destroyed	• Reduced profitability of larger-scale enterprises and increases in insurance premiums
Inland fishing operations and livelihoods	Changing levels of precipitation	Where rainfall decreases, reduced opportunities for farming, fishing and aquaculture as part of rural livelihood systems	• Reduced diversity of rural livelihoods, greater risks in agriculture • Greater reliance on non-farm income
	More droughts or floods	Damage to productive assets (fish ponds, weirs, rice fields, etc.) and homes	Increased vulnerability of riparian and floodplain households and communities
	Less predictable rainy/dry seasons	Decreased ability to plan seasonal livelihood activities, e.g. farming and fishing	

In developing countries, fish provides more than 20% of animal protein consumed, compared with just 8% in industrial countries. This percentage rises to more than 50% of total animal protein in small island developing states (World Bank, 2005).

Traditionally, island coastal communities in the Pacific have had some of the highest rates of fish consumption in the world – 3 or 4 times the global average – and they have relied upon fish to provide 50–90% of their dietary animal protein. Much of this fish comes from subsistence coastal fisheries based on coral reefs. By 2035, population growth is projected to reduce the availability of fish to below the 35 kg of fish per person per year recommended for good nutrition. In addition, the direct effects of ocean warming on fish metabolism, and the indirect effects of ocean warming on the quality of coral reef habitats as a result of coral bleaching and ocean acidification, are expected to reduce production by around 20% by 2050.

Table 2.11 shows the expected shortfall in coastal fisheries production based on coral reef area for four Pacific island nations (Laffoley and Baxter, 2016).

The rich tuna resources of the Pacific region offer a potential solution to this forecast deficit. But modelling of the effects of ocean warming on the most abundant tuna species in the region, skipjack tuna, indicates that there is likely to be an eastward shift in the relative abundance of this important species. Over time, it should be easier for coastal communities in Kiribati, the Cook Islands and French Polynesia to catch tuna than it is for coastal communities in the western Pacific (e.g. Papua New Guinea, and the Federated States of Micronesia).

In southeast Asia, the increase in marine and freshwater fish catch, and in the production from freshwater aquaculture and mariculture, has enabled per capita fish consumption to increase from about 13 to 32 kg per year since 1961. As a result,

Table 2.11 Coastal fisheries production and projected deficits for four Pacific island states.

Small island state	Coastal fish production, t/yr	Fish need for food (tonnes)		Surplus or deficit (tonnes)	
		2020	2035	2020	2035
Papua New Guinea	98,760	117,000	169,100	−18,200	−73,800
Solomon Islands	27,610	25,400	35,600	2210	−7990
Kiribati	12,960	10,900	13,400	2060	−890
Nauru	130	700	800	−570	−670

Source: Adapted from Laffoley and Baxter (2016). Reproduced with permission of International Union for Conservation of Nature.

present-day fish consumption in the region is well above the per capita global average of about 19 kg per year. Nevertheless, many fish stocks have been overexploited, resulting in fish abundances at the end of the twentieth century being 5–30% of the levels in the 1950s. Marine fisheries in the South China Sea are characterized by high numbers of fishing vessels collectively employing millions of people.

Maintaining the significant contribution of marine fisheries and mariculture to livelihoods in southeast Asia will be a huge challenge as the ocean continues to warm. Under a business-as-usual emissions scenario, harvests from marine fisheries are projected to decrease by 20–30% by 2050, relative to 1970–2000. Overall, a loss of more than 20% of the original fish species richness in southeast Asia seas is projected by 2050 (Laffoley and Baxter, 2016).

In the western Indian Ocean, the high population density in small island developing states relative to the coral reef area, coupled with poverty-driven dependence on fishing for food and cash, has caused the deterioration of fish habitats through overharvesting and destructive fishing. As a result, fisheries based on coral reefs now only make a modest contribution to per capita fish consumption, typically less than 5 kg/year for communities living within 25 km of the coast, although the Seychelles and Mayotte are the exception to this trend. By 2030, population growth coupled with the coral mortality linked to ocean warming is expected to reduce productivity even further (Laffoley and Baxter, 2016).

Food Insecurity

Food insecurity is a complex issue. It is defined by the FAO as: a situation where people lack secure access to sufficient amounts of safe and nutritional food for normal growth and development, and for an active and healthy life. It may be caused by the unavailability of food, insufficient purchasing power, inappropriate distribution, or inadequate use of food at the household level. Food insecurity, poor conditions of health and sanitation and inappropriate care and feeding practices are the major causes of poor nutritional status (FAO, 2014).

The converse of this situation, food security, is therefore the situation where people at all times have physical, social and economic access to sufficient, safe and nutritious food that meets their dietary needs and food preferences for an active and healthy life.

Based on this definition, four dimensions to food security can be identified:

- food availability;
- economic and physical access to food;
- food utilization;
- stability over time.

The *availability* dimension captures not only the quantity, but also the quality and diversity of food. Indicators for assessing availability include the adequacy of dietary energy supply; the share of calories derived from cereals, roots and tubers; the average protein supply; the average supply of animal-source proteins; and the average value of food production.

Access covers physical access and infrastructure such railway and road density; economic access, represented by the domestic food price index; and the prevalence of undernourishment.

Utilization has two components. The first includes factors that determine the ability to utilize food: access to water and sanitation. The second focuses on the outcomes of poor food utilization, for example, nutritional failure in children under five such as wasting, stunting and being underweight.

Stability also has two elements. The first covers factors that measure exposure to food security risk with a diverse set of indicators such as the cereal dependency ratio, the area under irrigation, and the value of staple food imports as a percentage of total merchandise exports. The second component focuses on the incidence of shock such as domestic food price volatility, fluctuations in domestic food supply, and political instability.

The different impacts of climate change on the various components of food security are shown schematically in Figure 2.7. Climate change has a direct impact not only on the production of food, but also on factors not directly related to production but which still strongly influence food security. (Porter *et al.*, 2014)

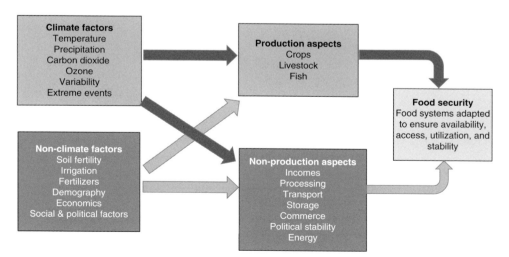

Figure 2.7 Climate and non-climate factors influencing food security. *Source:* Adapted from Porter *et al.* (2014).

The global impact of climate change on food security is generally assessed to be negative, but there are large variations across continents and regions. In a minority of cases (most of them in the higher latitudes), the net effect may even be positive, but for countries in the tropics (where all the SIDS lie), the impacts of climate change on food security are almost certain to be negative.

For the major crops (wheat, rice and maize) in both tropical and temperate regions, climate change will negatively impact production for local temperature increases of 2 °C or more, although individual locations may benefit. After 2050, the risk of more severe impacts increases, with projections consistently showing that production is negatively affected in tropical latitudes. Climate change will also progressively increase the inter-annual variability of crop yields in many regions.

Undernourishment in the SIDS

The primary manifestation of food insecurity is obviously hunger – defined as a chronic condition of undernourishment caused by the inability to acquire food intake that is sufficient to meet dietary requirements.

The FAO estimates published in 2014 showed that globally there has been continued progress in the reduction of hunger. About 805 million people were estimated to be chronically undernourished in 2012–14, a reduction of more than 100 million over the preceding decade, and more than 200 million less than 1990–92.

The decline in the share of undernourished and hungry people has been sharper than the decline in absolute numbers, because population gains have sometimes been greater than the reduction in hunger. Between 1992 and 2014, the level of undernourishment fell from 18.7% to 11.3% globally, and from 23.4% to 13.5% in developing countries.

Despite the progress in developing regions as a whole, there are large differences across regions. Africa has made slow progress – especially south of the Sahara, where one in four people remain undernourished.

The FAO does not report on undernourishment for all the SIDS. Those countries for which data were recorded are shown in Table 2.12 (FAO, 2014).

Apart from Haiti, all the SIDS on this list show progress in reducing undernourishment. But more than half the population of Haiti and more than a quarter of population of Timor-Leste remain chronically undernourished.

Poverty and Climate Change

Reducing poverty has been a key objective of the major international development agencies for decades. Progress has been slow but significant. But the changing climate is a major challenge to the eradication of poverty. Most of the shocks that keep households into poverty (or bring them to that level) are related to climate: natural disasters such as floods that destroy assets and disable families; health shocks such as malaria and diarrhoea (triggered or worsened by flooding) that result in increased expenditures on medicines and hospital charges; and lost labour income. Crop losses and food price

Table 2.12 Undernourishment in 20 small island states.

Small Island Developing States	Number undernourished, millions		Percentage of population undernourished	
	1990–92	2012–14	1990–92	2012–14
Barbados	<0.1	<0.1	<5.0	<5.0
Belize	<0.1	<0.1	9.7	6.7
Cabo Verde	<0.1	<0.1	16.1	9.9
Comoros	0.6	<0.04	5.7	<5.0
Dominican Republic	2.5	1.5	34.4	14.7
Fiji	<0.1	<0.04	6.6	<5.0
Guinea-Bissau	0.2	0.3	23.1	17.7
Guyana	0.2	<0.1	22.8	10.0
Haiti	4.4	5.3	61.1	51.8
Jamaica	0.2	0.2	10.4	7.9
Kiribati	<0.1	<0.005	7.5	<5.0
Maldives	<0.1	0.1	12.3	6.2
Samoa	<0.1	<0.009	10.7	<5.0
Sao Tome & Principe	<0.1	<0.1	22.9	6.8
Solomon Islands	<0.1	<0.1	24.8	12.5
St Vincent & Grenadines	<0.1	<0.1	20.7	5.7
Suriname	<0.1	<0.1	15.5	8.4
Timor-Leste	0.4	0.3	45.2	28.8
Trinidad & Tobago	0.2	0.1	12.6	9.0
Vanuatu	<0.1	<0.1	11.2	7.2

Source: Adapted from FAO (2014).

shocks due to drought, disease, or reduced yield due to temperature stress, are often catastrophic for poorer families.

Poor families are disproportionately affected because they are generally more exposed and more vulnerable to climate-related shocks, because they have fewer resources and receive less support from family, community, the financial system, and even social safety nets. Climate change will worsen these shocks and stresses, eventually decoupling economic growth and poverty reduction because of the constant shocks and setbacks caused by extreme weather, the increased frequency of natural disasters, and the inherent vulnerability of poorer communities (Hallegatte *et al.*, 2016).

Natural hazards, to which poor people are often more exposed and more vulnerable, are expected to become more intense and frequent in many regions. For instance, heatwaves that are considered exceptional today will become more common. In Europe, the 2003 heatwave, which led to more than 70,000 deaths, will be an average summer at the end of the century under a high-emission scenario.

The number of drought days could increase by more than 20% in most of the world by 2080, and the number of people exposed to droughts could increase by 9–17% in 2030 and 50–90% in 2080.

The number of people exposed to river floods could increase by 4–15% in 2030 and 12–20% in 2080, and coastal flooding risks can increase rapidly with sea-level rise.

In addition, poor people are strongly affected by diseases and health issues that climate change is likely to magnify (Hallegatte *et al.*, 2016).

Catastrophic events and natural disasters accelerate and intensify poverty because of the damage and loss they cause and because of the cost of rehabilitating the victims. They have three primary impacts:

1) **Loss of revenue by the poorest**: The repercussions of natural catastrophes are extremely grave for island economies because, in addition to causing losses and deaths, they provoke a considerable deterioration in revenue, in employment, and in resources when these events hit economic activities essential for the poor. These activities are agriculture, livestock raising, fishing, and other subsistence livelihoods. Moreover, the infrastructure and resources lost are not covered by any insurance. The poor are the hardest hit by these types of disasters because they tend to live and work in the zones where the risk is highest. The poor are the first to be decapitalized without any opportunity to recover or rebuild their assets.

2) **The necessity of urgency**: In the short term, the needs in terms of human and financial resources are enormous as government agencies try to respond to the urgency of rebuilding dwellings and damaged infrastructure. Rebuilding and rehabilitating impose substantial costs on the state and households, and above all on those that are not insured or where no government funds are allocated for these events. Funding originally allocated for investments are diverted to disaster relief. In calling for foreign assistance, countries increase their dependence, which in turn may undermine the local economy.

3) **Associated sickness:** A consequence of natural disasters is often the degradation of sanitation systems. The proliferation of bacteria and disease vectors together with the lack of clean water and the problem of evacuating wastewater and rainwater bring major health problems such as cholera. The poor have to take care of these problems themselves because they are rarely covered by insurance.

Tourism

Tourism is the mainstay of the economy for many small islands. Foreign currency receipts from international tourism often form a significant fraction of an island's gross domestic product. More than half of the SIDS count on international tourism as a major contribution to their economy. Table 2.13 shows the 13 SIDS where international tourism receipts account for more than 20% of GDP. In many islands, the number of international tourists arriving is significantly larger than the resident island population: testament to the substantial investments that have been made in tourism infrastructure in order to accommodate these visitors. Almost certainly, the majority of this infrastructure is on the coast and within the LECZ3 zone.

Table 2.13 The 13 SIDS where international tourism receipts account for more than 20% of GDP.

Name	Pop. 000s (2014)	International tourism arrivals, 2015 (000s)	International tourism receipts, M$ (2015)	GDP, M$	Int. tourism receipts as % of GDP
Anguilla	15	73	127	311	41%
Antigua and Barbuda	83	250	333	1248	27%
Aruba	104	1225	1652	2664	62%
Bahamas	393	1472	2379	8511	28%
Barbados	285	592	922	4353	21%
Belize	367	341	408	1699	24%
British Virgin Islands	31	393	484	902	54%
Dominica	73	74	128	533	24%
Maldives	370	1234	2567	3032	85%
Palau	22	162	127	234	54%
Seychelles	97	276	392	1511	26%
Saint Lucia	186	345	373	1406	27%
Vanuatu	271	90	257	812	32%

Sources: Adapted from UNDESA (2016); UNWTO (2016).

In the case of the Maldives, almost the entire economy is based on tourism, and more than a third of a million people depend on the industry. Yet the Maldives is one of the small island states most at risk from climate change.

The AIMS and Caribbean regions receive a greater number of visitors than the Pacific, but the ratio of visitor arrivals to the local population is growing sharply in the Pacific and AIMS regions, while remaining fairly steady in the Caribbean (UNDESA, 2014). However, there are large variations among the islands. The Bahamas, with a population of almost 400,000 people, accommodates as many as 6 million visitors a year. Providing electrical power and potable water for tourists is generally a major challenge for small island states.

Cruise ship tourism in the Caribbean is increasingly a large part of the regional industry. Visitors from cruise ships rose from 6 million in 2000 to 10.3 million in 2012, and represent about half of all international arrivals. Cruise ships are also becoming increasingly important in the Pacific. In 2013, for example, 133 cruise ships visited Vanuatu alone, with daily visits during some periods of the year (UNDESA, 2016). Cruise ship tourism presents a substantial environmental challenge: wastewater disposal, coastal degradation, fuel oil leakage, and an impact way over the carrying capacity of the local environment are major concerns.

The costs of future climate change impacts on coastal tourism are likely to be substantial. For instance, in the Caribbean a sea-level rise of just 1 metre is projected to put 266 out of 906 tourism resorts and 26 out of 73 airports in the Caribbean at risk of inundation. An estimated 49% of major tourism resorts in CARICOM (Caribbean

Community) would be damaged or destroyed by combined sea-level rise, storm surge and enhanced erosion, since many resorts lack any coastal protection, preferring to preserve the aesthetics of natural beach areas and views to the sea (Simpson *et al.*, 2011).

Freshwater Resources

Although the greatest threat to most small islands comes from the sea, many islands face another major problem: that of finding enough fresh water for urban populations, for industry, agriculture, and also for tourism. For the poorer islands where agriculture is a major part of the rural economy, any reduction in average rainfall, or changes in the pattern of rainfall over the year, could spell disaster for subsistence farmers.

In the tropics, all the surface flow comes from rainfall. Modelling the hydrological cycle and predicting how it may change over the course of the twenty-first century is mathematically difficult. Nevertheless, when the results of many computer simulations from a variety of different computer models are aggregated and analyzed, several clear trends emerge. Table 2.14 shows climate change projections for the intermediate Representative Concentration Pathway (RCP4.5) for the main SIDS regions. The simulation results are shown for the 25th, 50th, and 75th percentiles. The mean value is given by the 50th percentile (Nurse *et al.*, 2014).

All regions are expected to experience overall warming temperatures and a rise in mean sea-levels. But the changes in precipitation are much more variable. The projected changes in precipitation for the Caribbean and northern tropical Pacific are almost identical and signal a serious problem for the majority of the islands in this SIDS group. It should be noted that these trends accelerate moderately for RCP6.0 but steeply for RCP8 (AR5 WGII Section 29.4.3; Nurse *et al.*, 2014). In other words, these data may underestimate the expected decrease in precipitation for the Caribbean and Pacific SIDS.

Rainfall records across the Caribbean region for the period 1900–2000 already show a consistent reduction in rainfall – a trend that is likely to continue. In contrast, rainfall data over the same period for the Seychelles has shown a substantial variation related to the El Nino Southern Oscillation (Nurse *et al.*, 2014).

Table 2.14 Projected changes in temperature and precipitation for the main SIDS regions (RCP4.5 annual projected change for 2081–2100 compared with 1986–2005).

SIDS region	Temperature °C			Precipitation %			Sea-level rise, m
	25%	50%	75%	25%	50%	75%	
Caribbean	+1.3	+1.4	+1.9	−10	−5	−1	0.5–0.6
Northern tropical Pacific	+2.0	+2.3	+2.2	−10	−6	−3	0.4–0.5
Southern Pacific	+1.2	+1.4	+1.2	0	+1	+4	0.5–0.6
North Indian Ocean	+1.3	+1.5	+2.0	+5	+9	+20	0.4–0.5
West Indian Ocean	+1.2	+1.4	+1.8	0	+2	+5	0.5–0.6

Source: Adapted from Nurse *et al.* (2014). Reproduced with permission of the Intergovernmental Panel on Climate Change.

Human Health

Climate change will magnify many threats to health, especially for poor and vulnerable people, and particularly for children. The main diseases that affect poor people are diseases that are expected to worsen with climate change, such as malaria and diarrhoea.

Even modest temperature increases may significantly affect the transmission of malaria. Global warming of just 2 °C could increase the number of people at risk from malaria by up to 5%, or more than 150 million people (Hallegatte *et al.*, 2016). Climate change will bring new areas into risk from the disease – areas where control programmes are not yet in place and where people's natural immunity is weak. The incidence of diarrhoea is expected to worsen with a changing climate. Higher temperatures favour the development of pathogens, and the scarcity of water affects water quality and undermines good hygiene that would otherwise reduce the impact of the disease (WHO, 2003).

Small islands are often at greater risk from other diseases that flourish in warm climates: dengue, filariasis, leptospirosis and schistosomiasis. The Caribbean has been assessed as being a highly endemic zone for leptospirosis; Trinidad and Tobago, Barbados, Jamaica, Guadeloupe and Haiti all have high levels of the disease, as does the Seychelles.

The incidence of disease is often related to the climate, and seasonal variations are common, with rainy season impacts being higher. Warmer temperatures and increased flooding due to extreme weather are likely to increase the incidence of these impacts. Flooding in urban areas hugely increases the risk of cholera and typhoid. Cholera in Haiti killed thousands of islanders between 2010 and 2014 – a disease almost certainly started by UN peacekeepers sent to the island to help after the 2010 earthquake. Poor sanitation in the camps set up to shelter the tens of thousands of Haitians made homeless by the earthquake only made the problem worse. Hurricane Matthew in 2016 destroyed thousands of homes in southwest Haiti, and the resulting floods, poor drainage and lack of sanitation triggered a second wave of cholera.

In Pacific islands, the incidence of diseases such as malaria and dengue fever has been increasing, especially endemic dengue in Samoa, Tonga and Kiribati (Russell, 2009).

Dengue fever is a major health concern in Trinidad and Tobago, Singapore, Cape Verde, Comoros, and Mauritius (Nurse *et al.*, 2014), and in Trinidad and Singapore the outbreaks have been correlated with temperature and rainfall (Chadee *et al.*, 2005).

More recently, chikungunya has become widespread in the Caribbean. A mosquito-borne virus infection producing intense muscle pain, headache, nausea, and fatigue, it is a disease expected to become more widespread as climate change leads to warmer weather, and increased instances of urban flooding and sanitation systems that struggle to cope.

In 2005, a major outbreak of chikungunya occurred in many islands of the Indian Ocean. A large number of imported cases in Europe were associated with this outbreak, mostly in 2006 when the Indian Ocean epidemic was at its peak. An outbreak in India occurred in 2006 and 2007. Several other countries in southeast Asia were also affected. Since 2005, India, Indonesia, Maldives, Myanmar and Thailand have reported over 1.9 million cases of the disease. In 2007, transmission was reported for the first time in Europe, in a localized outbreak in northeastern Italy.

In December 2013, France reported two cases in the French part of the Caribbean island of St Martin. Since then, local transmission has been confirmed in over 43 countries and territories in the WHO Region of the Americas. As of April 2015, over

1.3 million suspected cases of chikungunya have been recorded in the Caribbean islands, Latin American countries, and the US; 191 deaths were attributed to this disease during the same period.

Since late 2014, outbreaks have been reported in the Pacific islands. In mid-2015, chikungunya was prevalent in the Cook Islands and the Marshall Islands, while the number of cases in American Samoa, French Polynesia, Kiribati and Samoa had diminished (WHO, 2017).

Ciguatera fish poisoning occurs in tropical regions and is the most common non-bacterial food-borne illness associated with eating fish. More than 50,000 cases of ciguatera poisoning occur globally every year. The illness is widespread in tropical and subtropical waters, and is particularly common in the Pacific and Indian Oceans and the Caribbean Sea.

Outbreaks of this illness correlate with water temperature, and there is concern that increasing temperatures associated with climate change could increase the incidence of ciguatera in the Caribbean, the Pacific and the Mediterranean (Nurse *et al.*, 2014). There are high rates in Tokelau, Tuvalu, Kiribati, the Cook Islands and Vanuatu (Chan *et al.*, 2011).

Ciguatera fish poisoning occurs after eating reef fish contaminated with toxins such as ciguatoxin or maitotoxin. These potent toxins originate from *Gambierdiscus toxicus*, a small marine organism (dinoflagellate) that grows on and around coral reefs and is ingested by herbivorous fish. The toxins are then modified and concentrated as they pass up the food chain to carnivorous fish and finally to humans. Since *G. toxicus* may proliferate on dead coral reefs more effectively than other dinoflagellates, the risk of ciguatera may increase as more coral reefs deteriorate because of climate change, ocean acidification, coastal pollution and nutrient runoff (CDC, 2015).

Fish that are most likely to cause ciguatera poisoning are carnivorous reef fish, including barracuda, grouper, moray eel, amberjack, sea bass or sturgeon. Omnivorous and herbivorous fish such as parrot fish, surgeonfish and red snapper can also be a risk.

Although the present incidence of malaria on small islands is not especially high, the disease is expected to spread slowly with the climate changes predicted for Papua New Guinea, Guyana, Suriname and French Guyana. In the Caribbean the disease is also expected to become more common on islands where currently the disease is not endemic (Nurse *et al.*, 2014).

Apart from these vector-borne diseases where the causality is easily established, climate change will have more insidious impacts on health. The poorer islands will suffer worse hardship as extreme weather and degraded ecosystems reduce the productivity of fisheries, marine ecosystems and agriculture. The poor will get poorer, and their health will suffer. Torrential rainfall and flooding in urban areas will overload and flood inadequate drainage and sanitation systems. Bacterial diarrhoea and cholera will become endemic on many small islands.

Climate-driven Migration

Faced with these threats, many islanders are contemplating moving to higher ground, particularly in the Pacific, where sea-level rise has been four times higher than the global average. On many Pacific islands, low-lying coastal lands will be overwashed and flooded so often that they will soon become uninhabitable. The migration has already begun.

Health and disease in the SIDS

Cardiovascular disease is the leading cause of death in the Pacific islands. In 2012, the World Bank reported than non-communicable diseases (NCDs) accounted for 70% of all death in 12 Pacific island countries. The Caribbean is also afflicted: the region suffers from the highest prevalence of chronic NCDs in the Americas. Heart disease and cancer are major threats among both males and females, and the economic burden of NCDs on Caribbean countries has been increasing exponentially since 2010. In all three regions, chronic diseases lead to considerable loss of productivity and take up a large fraction of healthcare resources available for diagnosis, treatment and rehabilitation.

Funding allocated to public health in the SIDS has been increasing. The Pacific region has the largest relative expenditure on public health: in 2010 expenditures were almost 8% of GDP compared with half that in the Caribbean and a third of that amount in the AIMS region (2.8%). SIDS populations do not have access to a wide range of much-needed healthcare resources such as primary-care clinics, specialist centres and tools for promoting healthy lifestyles. SIDS also suffer from a lack of qualified physicians and healthcare professionals.

Source: UNDESA (2014).

In the Solomon Islands, the town of Taro, the regional capital of Choiseul Province, is to be relocated to higher ground on an adjacent island. Dozens of villages on Fiji are to be moved, and 2000 people from the Carteret atoll on Papua New Guinea are likely to be moved to mainland Bougainvillea (Albert *et al.*, 2016).

In 2014, the Kiribati government bought $20\,km^2$ of land on Vanua Levu, one of the Fiji islands, in case its people cannot be moved internally. Kiribati has a policy called 'migration with dignity' if its archipelago of 33 coral atolls becomes uninhabitable.

On the Maldive islands, it is planned that the Malé commercial port will be moved to another island, Thilafushi, a location that is reckoned to be better protected against sea-level rise.

This is just the tip of the iceberg; the UNHCR estimates that climate change could force the displacement of more than 200 million people over the next 30 years. This level of migration will inevitably lead to extensive and chronic regional conflicts as countries struggle to cope with hundreds of thousands of migrants and climate refugees, many of whom will try to enter countries illegally in a desperate attempt to find safety and security for themselves and their families. There can be little doubt that climate change will eventually create massive social problems for practically *all* countries.

References

Albert, S., Leon, J.X., Grinham, A.R., *et al.* (2016) Interactions between sea-level rise and wave exposure on reef island dynamics in the Solomon islands. *Environmental Research Letters*, 11, 054011.

Beebe, W. (1928) *Beneath Tropical Seas: A Record of Diving among the Coral Reefs of Haiti.* NY Zoological Society, Putnam and Sons, New York.

Burke, L., Reytar, K., Spalding, M. and Perry, A. (2011) *Reefs at Risk Revisited.* World Resources Institute, Washington, DC.

Burke, L., Reytar, K., Spalding, M. and Perry, A. (2012) *Reefs at Risk Revisited in the Coral Triangle*. World Resources Institute, Washington, DC.

CDC (2015) Food poisoning from marine toxins. Centers for Disease Control and Prevention. Available at wwwnc.cdc.gov/travel/yellowbook/2016/the-pre-travel-consultation/food-poisoning-from-marine-toxins

CEPF (2010) *Critical Ecosystem Partnership Fund: Ecosystem Profile. The Carribbean Islands Biodiversity Hotspots*. Critical Ecosystem Partnership Fund, Arlington, VA.

Chadee, D.D., Williams, F.L.R. and Kitron, U.D. (2005) Impact of vector control on a dengue fever outbreak in Trinidad, West Indies, in 1998. *Tropical Medicine & International Health*, 10(8), 748–758.

Chan, W.H., Mak, Y.L., Wu, J.J., *et al.* (2011) Spatial distribution of ciguateric fish in the Republic of Kiribati. *Chemosphere*, 84(1), 117–123.

CIESIN (2013) *Low Elevation Coastal Zone (LECZ) Urban-Rural Population and Land Area Estimates, Version 2*. Center for International Earth Science Information Network, CIESIN, Columbia University. Available at http://dx.doi.org/10.7927/H4MW2F2J. NASA Socioeconomic Data and Applications Center (SEDAC), Palisades, NY.

Dai, A., Qian, T. and Trenberth, K.E. (2008) Changes in continental freshwater discharge from 1948 to 2004. *Journal of Climate*, 22, 2773–2792.

FAO (2007) *Building Adaptive Capacity to Climate Change. Policies to Sustain Livelihoods and Fisheries*. New Directions in Fisheries, No. 8. Food and Agricultural Organization, Rome.

FAO (2013) *FAO Statistical Yearbook 2013*. Food and Agricultural Organization, Rome.

FAO (2014) *The state of food insecurity in the world: Strengthening the enabling environment for food security and nutrition*. Food and Agricultural Organization, Rome.

Hallegatte, S., Bangalore, M., Bonzanigo, L., *et al.* (2016) *Shock Waves: Managing the Impacts of Climate Change on Poverty*. Climate Change and Poverty Series. World Bank, Washington, DC.

HDR (2014) *Human Development Report 2014*. World Bank, Washington, DC.

Hoeke, R.K., McInnes, K.L., Kruger, J.C., *et al.* (2013) Widespread inundation of Pacific islands by distant-source wind-waves. *Global Environmental Change*, 108, 128–138.

Kumar, L. and Taylor, S. (2015) Exposure of coastal built assets in the South Pacific to climate risks. *Nature Climate Change*, 5, 992–996. doi:10.1038/nclimate2702

Laffoley, D. and Baxter, J.M. (eds) (2016) *Explaining Ocean Warming: Causes, Scale, Effects and Consequences*. International Union for the Conservation of Nature, Gland, Switzerland.

Miller, M., McClennan, D.B. and Wiener, J.W. (2007) Apparent rapid fisheries escalation at a remote Caribbean island. *Environmental Conservation*, 34(2), 92–94.

National Hurricane Center (2015a) Storm surge overview. National Hurricane Center, NOAA. Available at www.nhc.noaa.gov/surge/?text

National Hurricane Center (2015b) Hurricanes in history. National Hurricane Center, NOAA. Available at www.nhc.noaa.gov/outreach/history

National Hurricane Center (2016) National Hurricane Centre. Hurricane Matthew Public advisory No. 25, 4 October 2016.

Nurse, L.A., McLean, R.F., *et al.* (2014) *IPCC 5th Assessment Report, Chapter 29. Small Islands*. Intergovernmental Panel on Climate Change, Cambridge.

Porter, J.R., Xie, L., *et al.* (2014) Food security and food production systems, in *Climate Change 2014: Impacts, Adaptation and Vulnerability. Part A: Global and Sectoral*

Aspects. Contribution of Working Group II to the Fifth Assessment Report of the Intergovernmental Panel on Climate Change. Cambridge, Cambridge University Press.

Reef Check (2016) *Where to Designate Marine Protected Areas in Haiti?* Reef Check, California, USA. See www.reefcheck.org for information about Reef Check's work worldwide.

Russell, L. (2009) *Climate Change and Health in Pacific Island Countries*. Centre for Health Policy, University of Sydney, Australia.

Simpson, M., Scott, D. and Trotz, U. (2011) *Climate Change's Impact on the Caribbean's Ability to Sustain Tourism, Natural Assets, and Livelihoods*. Inter American Development Bank, Miami.

The Lancet (2015a) *Health and climate change: policy responses to protect public health*. The Lancet Commissions. doi:10.1016/S0140-6736(15)60854-6

The Lancet (2015b) Adapting to migration as a planetary force. Editorial published online. *The Lancet*, 386, September 12.

UN (2008) *Handbook on the Least Developed Country category: Inclusion, graduation and special support measures*. UN Department of Economic and Social Affairs, New York.

UN (2011) *Small Island Developing States: Small Islands, Big(ger) Stakes*. Office of the High Representative for the Least Developed Countries, Landlocked Developing Countries and Small Island Developing States, UN-OHRLLS. New York.

UN (2012) Economic vulnerability indices. Available at www.un.org/en/development/desa/policy/cdp/ldc/excel/2012_ldc_data.xls

UN (2015) List of least developed countries. Available at www.un.org/en/development/desa/policy/cdp/ldc/ldc_list.pdf

UNDESA (2014) *Trends in Sustainable Development: Small Island Developing States (SIDS)*. United Nations Department of Economic and Social Affairs, New York.

UNDESA (2016) *World Statistics Pocketbook 2016 Edition*, Series V, No. 40. United Nations Department of Economic and Social Affairs, New York.

UNEP (2015) *The Emissions Gap Report 2014*. United Nations Environment Program (UNEP), Nairobi.

UNWTO (2016) *Tourism Highlights 2016*. World Tourism Organisation, Madrid.

USAID (2016) *USAID Climate Action Review: 2010–2016*. United States Agency for International Development, Washington, DC.

WHO (2003) *Summary Booklet: Climate Change and Human Health – Risks and Responses*. World Health Organization, Geneva.

WHO (2017) Available at http://www.who.int/mediacentre/factsheets/fs327/en/

WMO (2014) The escalating impacts of climate-related natural disasters. WMO Report July 2014. World Meteorological Office, Geneva.

Wong, P.P., Losada, I.J., Gattuso, J.-P., *et al.* (2014) Coastal systems and low-lying areas, in *Climate Change 2014: Impacts, Adaptation and Vulnerability. Part A: Global and Sectoral Aspects*. Contribution of Working Group II to the Fifth Assessment Report of the Intergovernmental Panel on Climate Change. Cambridge, Cambridge University Press, pp. 361–409.

World Bank (2005) *Turning the Tide: Saving Fish and Fishers*. World Bank, Washington, DC.

3

Adapting to a Changing Climate

This book is primarily about adaptation to climate change, not mitigation, for the simple reason that if all the Small Island Developing States were to reduce their greenhouse gas emissions to zero tomorrow, the impact on global concentrations of carbon dioxide would be negligible. In a full year, the emissions of all the 51 SIDS combined only add up to what the industrial countries together emit in about three days. This is not to say that mitigation measures should be ignored, but for the SIDS, mitigation is not the priority.

However, there are important and effective adaptation measures that also substantially reduce net emissions of carbon dioxide. Increasing the areas of forested land is the obvious example, where the many advantages and benefits of perennial vegetative ground cover, particularly on steep eroding slopes, include the capture and sequestering of carbon dioxide. This logic also applies to increasing the area of coastal mangroves, where these trees protect coastlines, reduce erosion, and provide a safe haven and breeding ground for large numbers of fish and other marine life in coastal waters. The traditional cultivation of rice paddy produces substantial quantities of methane; cultivation that uses water more efficiently emits less methane. Adapting to climate change by using water more efficiently and switching to different forms of cultivation such as conservation agriculture makes a significant contribution to reducing greenhouse gases.

In addition, there are measures that fall more under the heading of sustainable development that make sound economic sense, and which will also reduce carbon emissions. Investing in renewable energy, improving energy efficiency, and transitioning to a low-carbon economy should be part of development planning for all the small island states. Not only will emissions of greenhouse gases be reduced, but island economies will become naturally more robust and resilient.

The previous chapter has outlined the major threats from climate change to the 51 small island states. We have argued that the main threats come from the sea: first from storm surges and flooding driven by tropical storms and violent weather exacerbated by rising sea-levels; and secondly, and more subtly, from the increasing deterioration of coral reef systems and their symbiotic coastal ecosystems which threatens to undermine and weaken the economies of all the island states that depend on fisheries and coastal tourism. Stronger and more resilient marine ecosystems also mean more fish: a substantial economic benefit for island populations and their tourist visitors.

The third adaptive response focuses on managing freshwater resources. Rainwater must be recognized as a valuable resource that cannot be wasted or squandered – except

Climate Change Adaptation in Small Island Developing States, First Edition. Martin J. Bush.
© 2018 John Wiley & Sons Ltd. Published 2018 by John Wiley & Sons Ltd.

when there is so much of it that it threatens to flood and destroy the land that depends on it for agriculture and water supply. So rainwater must be captured, collected, stored and much more carefully managed. And here there is a resource that is often overlooked and yet which has the potential to make an important impact: this is the huge potential of recycled wastewater used for irrigated agriculture.

Finally, agricultural production and food security are everywhere threatened by climate change. Much can be done to adapt to higher temperatures, more frequent droughts, variable precipitation and extreme weather.

Ecosystem-based Adaptation

Ecosystem-based adaptation (EBA) integrates the management of biodiversity and ecosystem services into an overall strategy to help communities adapt to the adverse impacts of climate change. It focuses on the sustainable management, conservation and restoration of ecosystems to provide services that help people and communities adapt to both present climate variability and longer-term climate change.

There are five key components to the EBA approach (Colis *et al.*, 2009):

- **Biodiversity conservation**: Protecting, restoring and managing key ecosystems helps biodiversity and people to adjust to changing climatic conditions. Ecosystem-based adaptation can safeguard and enhance protected areas and fragile ecosystems. It can also involve the restoration of fragmented or degraded ecosystems and the rehabilitation of weakened ecosystem processes such as migration or pollination.
- **Sustainable water management**: Managing, restoring and protecting ecosystems can contribute to sustainable water management by improving water quality, increasing groundwater recharge and reducing surface water runoff during storms. About a third of the world's largest cities obtain a significant proportion of their drinking water from forested protected areas.
- **Livelihood sustenance and food security**: By protecting and restoring healthy ecosystems to be more resilient to climate change impacts, ecosystem-based adaptation strategies can help to ensure continued availability and access to essential natural resources so that communities can better cope with climate variability and future climate change. In this context, EBA can directly meet the needs of community-based adaptation and poverty reduction initiatives.
- **Disaster risk reduction**: Ecosystem-based adaptation measures frequently complement disaster risk reduction objectives. Healthy ecosystems play an important role in protecting infrastructure and enhancing human security, acting as natural barriers and mitigating the impact and aiding recovery from many extreme weather events such as coastal and inland flooding, droughts, extreme temperatures, fires, landslides, hurricanes and cyclones.
- **Carbon sequestration**: Ecosystem-based adaptation strategies can complement and enhance climate change mitigation. Sustainable management of forests can store and sequester carbon by improving overall forest health and simultaneously sustain functioning ecosystems that provide food, fibre and water resources that people depend upon. Conservation and, in some cases, restoration of peatlands and wetlands can protect very significant carbon stores.

One of the key advantages of the EBA approach is that restoring and strengthening ecosystems *always* generates associated benefits, because the services provided by eco-systems are multidimensional. Examples of EBA measures that provide co-benefits are outlined in Table 3.1.

Adaptation in Coastal and Marine Environments

Coastal societies and ecosystems are particularly vulnerable to climate change. Even a 1 metre rise in sea-level could displace nearly six million people across southeast Asia, and 37 million more along the river deltas of east Asia (Dasgupta *et al.*, 2007). Protection strategies for sea-level rise range from 'hard' defences, such as sea walls, dykes and tidal barriers, to 'soft' defences such as strengthening and extending natural ecosystems such as mangroves, wetlands, mudflats and sand-dunes. In most developed countries, hard structures have been the preferred option particularly in builtup areas. However, these defences have often been built with little regard for the integrated nature of coastal ecosystems, and they frequently require costly repairs and upgrades (Spalding *et al.*, 2014). In some cases, hard structures can actually reduce the adaptation potential of coastal ecosystems, a situation characterized as maladaptation. More recent strategies include coastal retreat, where infrastructure is moved inland to reduce the risk of impacts and to allow the development of intertidal ecosystems, or planning restrictions that prevent the development of infrastructure on floodplains or at a certain distance from the shore.

Resilient ecosystems are an effective partner in defending and protecting coastal zones. Soft solutions include the planting of marsh vegetation in the intertidal zone, and restoring mangrove forests and wetlands. Coastal wetlands absorb wave energy and reduce erosion through increased drag on the movement of water, and a reduction in the direct effects of wind. The accretion of sediments helps to maintain shallow depths that reduce the strength of waves.

Ecosystem-based adaptation measures are receiving increased attention in developing countries, particularly in the SIDS, where adaptive capacity is often low and local communities depend much more on their natural resource base. Mangroves, for example, provide physical protection to coastal communities while providing benefits such as productive fisheries, offering both physical protection and economic gain to the most vulnerable communities. Nearly 12,000 hectares of mangroves planted in Vietnam at a cost of 1.1 M$ saved an estimated 7.3 M$ per year in dyke maintenance while providing protection against a typhoon that devastated neighbouring areas (Reid and Huq, 2005). In the US, coastal wetlands are estimated to provide over $23 billion per year in hurricane protection (Costanza *et al.*, 2008).

Some studies suggest that it is preferable to use expensive structural protection in highly developed areas, and softer approaches such as land management in less developed areas (Kirshen *et al.*, 2008). In addition to being more cost-effective, strategies focused on ecosystem restoration and management generally provide co-benefits as well as contributing to mitigation through carbon sequestration.

Effective coastal zone management requires an integrated approach because of the interconnectedness of coastal ecosystems. Waves approaching a coast travel across

Table 3.1 Co-benefits of ecosystem-based adaptation measures.

		Co-benefits			
Adaptation measure	**Adaptation function**	**Social and cultural**	**Economic**	**Biodiversity**	**Mitigation**
Coral reef sustainable management	Protects against storm surge and coastal erosion	Culture of reef fishing sustained and enhanced	Generation of higher incomes from fishing, increased food security	Conservation of marine ecosystems and diversification of fish species	No change
Mangrove forest restoration and conservation	Protection against storm surge, sea-level rise, erosion and coastal inundation	Provision of employment options (fisheries and crustaceans), contribution to food security	Generation of income for local communities through marketing of mangrove products (fish, crustaceans, dyes, medicines)	Conservation of species that live or breed in mangroves	Conservation of carbon stocks above and below ground
Forest conservation and sustainable forest management	Maintenance of nutrient and freshwater flows, stabilization of eroded slopes, prevention of landslides	Recreation and tourism, protection of indigenous peoples and local communities, improved microclimate	Generation of income through ecotourism, recreation, sustainable logging and forest plants	Conservation of habitat for forest plants and animal species	Conservation of carbon stocks, reduction of emissions from deforestation and degradation
Restoration of degraded wetlands	Maintenance of nutrients and freshwater flow, water quality, storage and capacity, protection against floods or storm inundation	Sustained provision of livelihoods, recreation, and employment opportunities	Increased livelihood options, potential revenue from recreational activities, sustainable use	Conservation of wetland flora and fauna through maintenance of breeding grounds and stopover sites for migratory birds	Reduced emissions from soil carbon mineralization
Establishment of diverse agroforestry systems in agricultural land	Diversification of agricultural production to cope with changed climatic conditions	Contribution to food and fuel wood security	Generation of income from sale of timber, firewood and other products	Conservation of biodiversity in agricultural landscapes	Carbon storage in above-ground and below-ground biomass and soils

Conservation of agrobiodiversity	Provision of specific gene pools for crop and livestock adaptation to climatic variability	Enhanced food security, diversification of food products, conservation of local and traditional knowledge and practices	Possibility of agricultural income in difficult environments, environmental services such as bees for pollination of cultivated crops	Conservation of genetic diversity of crop varieties and livestock breeds	No change
Conservation of medicinal plants used by local and indigenous communities	Local medicines available for health problems resulting from climate change or habitat degradation, e.g. malaria, diarrhoea, cardiovascular problems	Local communities have an independent and sustainable source of medicines, maintenance of local knowledge and traditions	Potential sources of income for local people	Enhanced medicinal plant conservation, local and traditional knowledge recognized and protected	No change
Sustainable management of grasslands	Protection against floods, storage of nutrients, maintenance of soil structure	Recreation and tourism	Generates income for local communities through production from plants (fibre, etc.)	Forage for grazing animals, provides diverse habitats for animals that are predators and prey	Maintenance of soil carbon storage

Source: Courtesy of Convention on Biological Diversity (CBD, 2009).

reefs and through seagrass beds before reaching mangroves, and the power of the wave is not dissipated by one ecosystem alone. One estimate is that a 100-metre wide belt of mangrove forest can reduce wave intensity by 90% (Spalding *et al.*, 2014). However, many coastal and marine ecosystems are becoming increasingly degraded. Although climate change could result in a 10–15% loss of mangroves, the current rate of deforestation far exceeds this level. This is particularly true in Haiti. Mangroves also respond to sea-level rise and coastal erosion by retreating inland, and may be more severely impacted if there is a reduced area to move into. Land-use planning is necessary to take this phenomenon into account and to avoid what is called coastal 'squeeze' (Gilman *et al.*, 2008).

Several studies carried out following coastal disasters such as tsunamis and hurricanes have documented the important role of wetlands, mangroves and coral reefs in coastal protection against extreme events and tropical storms. An assessment following the massive tsunami of 2004 found a clear correlation between damage on inland areas and human modifications to the coastline, with mature sand-dunes especially effective in providing protection against storm surges (Alongi, 2008). During Hurricane Katrina, levees formed by extensive wetlands escaped substantial damage, suggesting that a well-managed combination of hard and soft protection can play an effective role in climate change adaptation and that the re-establishment of protective habitats could be important even for builtup areas.

However, it is unlikely that coastal and marine ecosystems alone can fully attenuate the impacts of storms. A combination of soft and hard defences coupled with more effective community organization is essential if the impact of extreme weather is to be reduced to tolerable and manageable levels. In addition, the other factors that weaken the resilience of coastal and marine ecosystems must be included in an integrated approach. These factors include urban and marine pollution, agricultural nutrient run-off, excessive coastal sedimentation from the erosion of watersheds, untreated sewage from coastal urban areas, and unsustainable tourism. This is sometimes called a 'ridge-to-reef' approach to natural resources management in coastal environments. But while useful as a reminder of the spatial extent of the analysis required, it needs to explicitly include the stress factors caused by careless urban planning, inefficient agriculture, eroded watersheds and poorly managed beach tourism.

Approaches to adapting to sea-level rise and coastal erosion can be broadly classified as *protect*, *accommodate*, or *retreat*. These alternatives are outlined in Table 3.2 (Linham and Nicholls, 2010).

Most countries will invest first in protection measures, particularly to protect the shoreline, coastal infrastructure and beach tourism. Here, there is an ongoing debate about the relative merits of 'hard' and 'soft' defences.

Hard Defences

Building sea walls is the traditional response to beach erosion and occasional flooding. The approach is based on installing a solid barrier between the land and the sea in order to resist the energy of tides and waves. This installation prevents any interaction between the land and the sea. Examples include seawalls, sea dykes, revetments and breakwaters. Hard defences have been employed because they provide tangible and visible protection. Over time, however, it has become clear that while these structures

Table 3.2 Responses to sea-level rise.

	Protect	Accommodate	Retreat
Built environment	Protect coastal development, e.g. seawalls, dykes, beach nourishment, sand dunes, surge barriers, land claim	Regulate building development and increase awareness of hazards, e.g. flood-proofing, flood hazard maps, flood warnings	Establish building setback codes, e.g. managed realignment, coastal setbacks
Wetlands	Create or restore wetland habitats and mangrove forests by land-filling and planting	Strike balance between preservation and development	Allow wetland and mangrove migration, e.g. managed realignment, coastal setbacks
Agriculture	Protect agricultural land with sea walls, dykes, beach nourishment, sand dunes, surge barriers, and land claims	Switch to aquaculture or floating agriculture	Relocate agricultural production, e.g. managed realignment, coastal setbacks

Source: Adapted from Linham and Nicholls (2010).

provide benefits to the hinterland they protect, they do little to prevent the physical process of erosion. Instead the problem is transferred from the shoreline to the seabed immediately in front of the structure or the adjacent coast.

Once constructed, hard defences fix the location of the coastline in the position at the time of construction. Although this may well be the objective, fixing the position of a coastline is a problem because coastlines are naturally dynamic landforms that respond to factors such as rising sea-levels and changing wave action. Hard structures may also impede the recreational use of beaches and can be costly to construct and maintain.

Increasing awareness of the negative side-effects of hard structures on erosion and sedimentation patterns has led to growing recognition of the benefits of 'soft' protection and the adaptation strategies of accommodation and retreat. Alternatively, hard defences can be combined with soft defences such as beach nourishment. Using this approach, nourishment would maintain beach levels while the hard defences continue to protect the coastline against the most extreme events.

Soft Defences

While hard defences fight against natural forces such as wave energy, soft technologies adapt to and supplement natural processes. The move towards soft defences has largely been driven by a response to the negative impact of hard defences. Soft defences also represent a more integrated response to the threat of coastal erosion.

The introduction of soft defences helps maintain the natural landscape and habitat functionality of the coast. By working with (not against) natural processes, there is also an increased potential for maximizing the benefits of a scheme while minimizing the environmental impacts.

Soft defences also require regular monitoring, maintenance and engineering (in the case of beach nourishment).

It is probable that both types of defence will continue to be used. Hard defences are more likely to be important in protecting coastal urban areas against sea-level rise and flooding. Soft defences are more likely to be used on sedimentary coasts composed of beaches. In some places, it may be advantageous to use both hard and soft defences, where the hard structures are set back a short distance from the shoreline and constitute a barrier of 'last resort'.

Marine Protected Areas

There are now thousands of protected areas and national parks, including marine parks, around the world. A protected area is a clearly defined geographical space dedicated and managed, through legal or other effective means, to achieve the long-term conservation of nature together with ecosystem services and cultural values (IUCN, 2014).

Protected areas – national parks, wilderness areas, community conserved areas, and nature reserves – are the cornerstone of efforts to ensure the conservation of biodiversity, while also contributing to people's livelihoods, particularly at the local level. Protected areas provide important services: food, clean water supply, medicines, and protection from the impacts of natural disasters. Their role in helping mitigate and adapt to climate change is also increasingly recognized; it has been estimated that the global network of protected areas stores at least 15% of terrestrial carbon.

A marine protected area (MPA) is a geographically defined area, legally established, that is regulated and managed to achieve specific conservation objectives. A MPA generally includes several levels of protection and zones that are regulated and managed in different ways. The strictest level is a 'no-use zone' where no fishing or extraction of marine resources is allowed. It is simply off-limits. A 'no-take zone' allows access to divers and swimmers, but all extractive activities (including fishing) are excluded. There are also intermediate peripheral zones, called buffer zones, which are more lightly regulated but still monitored and supervised by park personnel. Multiple use areas allow all types of recreation, fishing and diving, but these areas too must be monitored and protected.

Marine protected areas have been shown to provide multiple benefits to local and coastal communities. The restrictions on fishing allow fish populations to recover and flourish, and the amount of fish caught outside no-take zones generally more than compensates for the initial prohibition. Coral reefs recover their functionality and attract more fish. Mangrove forests shelter juvenile species and contribute to the productivity of the protected area.

The past two decades have seen a dramatic increase in the number of MPAs within the SIDS exclusive economic zones (EEZ). The area of MPAs established in the 51 SIDS reached $70,200 \, km^2$ in 2012 (Figure 3.1), more than eight times the total area protected in 1990 (UNDESA, 2014). These numbers do not include the huge Phoenix Islands Protected Area in Kiribati, which alone covers more than $408,000 \, km^2$.

Marine Park Management

It is, however, legally relatively easy to establish a protected area, but rather more difficult to manage it effectively. Protected areas that exist only on paper are called 'paper parks' and unfortunately many protected areas and marine parks warrant the name.

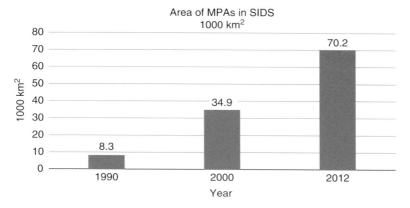

Figure 3.1 Marine protected areas in small island developing states.

The first clue as to whether a protected area is a serious proposition is to check out the management plan. A park without a management plan is like a ship without a rudder. It may be moving forward but nobody quite knows where it is going or how to get it to its supposed destination. And parks without a management plan soon grind to a halt. What was once supposed to be a protected area no longer is, and the area soon becomes exploited as before: overfished, unmanaged and unprotected.

A management plan is essential. It specifies the objectives – what the MPA is intended to achieve – and sets out how the area is to be managed so as to achieve these objectives. There will generally be more than one objective, and so the area may be divided into different zones each of which is managed in a different way.

Table 3.3 outlines a typical MPA site management plan. It need not be a substantial document. The idea is to outline and summarize the management procedures required – this should not require too much text.

Managed areas that protect and restore ecosystems with the objective of allowing their sustainable use are designated by the IUCN as Category VI protected areas. Areas that are managed with the aim of restoring coastal and marine ecosystems and strengthening their resilience to climate change impacts on coastal areas are also Category VI protected areas. However, there are important differences as outlined in Table 3.4.

One important difference is that EBA managed areas on small islands are areas that should *always* cover both the marine and the coastal environments including the terrestrial ecosystems that impact the coastal zone, such as the watersheds whose watercourses discharge along the coast. The long-term resilience of coastal areas requires comprehensive approaches that identify and address risk factors within the broader coastal landscape and catchment area. Improved watershed management upstream, and the reforestation and restoration of degraded lands, will reduce excessive sediment loads on downstream coastal waterways, and areas that cause siltation of natural ponds, mud flats, and wetland systems, as well as in some places contributing to coastal accretion. Coherent plans for the restoration and protection of such natural systems within a broader landscape approach are essential for increasing coastal resilience and successfully adapting to climate change.

As an example of the ecosystem-based adaptation approach applied within a managed marine area, we look at the Three Bays Protected Area in Haiti.

Table 3.3 Typical MPA management plan.

Typical marine protected area management plan

Executive summary

Purpose and scope of the plan

Legislative authority for the park

Map of the area and the different management zones

Main report

1 Site location and situational analysis

– Reefs, beaches, wetlands, intertidal zone, sea grasses, mangroves

– Biodiversity: marine, terrestrial and ornithological

– Critical habitats for endangered species

– Cultural, historical, and archeological points of importance

2 Existing uses

– Recreational

– Commercial (including artisanal)

– Research and education

– Traditional user rights and management practices

– Destructive practices (dynamite, poisoning, extraction, overfishing)

3 Community structure and involvement

4 Local government structure and involvement

5 Existing legal and management framework

6 Threats: present and potential and how they should be resolved and managed. Expected climate change impacts on the region and their impact on the MPA

7 Management procedures

– Goal and objectives

– Boundaries, zoning, and regulations

– Management structure: advisory committee, stakeholder participation, co-management

– Community participation

– Social and cultural aspects

– Interpretation plan

– Infrastructure required (visitor centre, boat mooring areas, diving facilities, etc.)

8 Administration

– Staffing requirements

– Training

– Facilities and equipment

9 Surveillance, monitoring and enforcement

10 Monitoring and evaluation

– Performance indicators and performance management plan

11 Financial management

– Budget

– Funding sources and private sector involvement

– Income generating activities

– Business plan

Source: Adapted from Salm *et al.* (2000). Reproduced with permission of the International Union for Conservation of Nature.

Table 3.4 MPA objectives in the context of climate change.

IUCN Category VI protected area objectives	MPA adaptation objectives
To promote sustainable use of natural resources, considering ecological, economic and social dimensions.	To restore, strengthen, and build resilience to climate change in marine and coastal ecosystems, and manage their sustainable use employing an integrated approach that includes the terrestrial environment.
To promote social and economic benefits to local communities where relevant.	To restore and strengthen coastal and marine ecosystem services and provide social, cultural, and economic benefits to local communities through a co-management approach.
To facilitate intergenerational security for local communities' livelihoods – therefore ensuring that such livelihoods are sustainable.	Restoring ecosystems and ensuring that development is carbon-neutral and sustainable.
To integrate other cultural approaches, belief systems and world-views within a range of social and economic approaches to nature conservation.	A secondary objective.
To contribute to developing and/or maintaining a more balanced relationship between humans and the rest of nature.	To contribute to developing a balanced relationship between communities and the ecosystems they depend on.
To contribute to sustainable development at national, regional and local level (in the last case mainly to local communities and/or indigenous peoples depending on the protected natural resources).	To contribute to sustainable development at national, regional (subnational), and local levels by ensuring that ecosystem services are restored, strengthened, and sustainable.
To facilitate scientific research and environmental monitoring, mainly related to the conservation and sustainable use of natural resources.	Scientific research is not a priority. However, monitoring of ecosystem health and functionality must be regularly conducted and management strategies adapted as needed.
To collaborate in the delivery of benefits to people, mostly local communities, living in or near to the designated protected area.	To ensure that coastal communities are better protected from climate change impacts, particularly sea-level rise, storm surges, and coastal flooding.
To facilitate recreation and appropriate small-scale tourism.	To manage recreation and tourism so that it is sustainable, and generates revenues that contribute to the cost of managing the area.
	To manage and reduce collateral risk factors from the terrestrial coastal environment: pollution, eutrophication, sedimentation, erosion, watershed flooding.

Three Bays Protected Area in Haiti

The Three Bays Protected Area (commonly referred to as PN3B) was gazetted in March 2014, appearing in Haiti's official government journal, *Le Moniteur*, as 'Aire protégée de ressources naturelles gerées des trois baies.'[1] The Three Bays are situated along

---

1 The name translates as 'Protected area for managed natural resources of the Three Bays'.

the northeast coast of Haiti between the town of Cap Haitien and the border with the Dominican Republic (DR).

The bays are named after the three towns that front the coastal areas: Limonade, Caracol and Fort LIberté (see Figure 3.2). Also included is an important wetland area adjacent to the DR border called Lagon aux Boeufs. The western boundary is defined by a line that starts 30 metres from the eastern edge of the Grand Riviere du Nord, and runs directly north to Haiti's territorial limit; the park boundary is then traced due east for 2 km, before angling down to the southeast to join the coast at the DR border. The terrestrial limit follows the 10-metre contour except in the area of Fort Liberté where it is defined by the main road running east–west along the coast.

Ideally, the PN3B protected area would include more of the coastal area, including the lower parts of the watersheds of the main rivers that discharge into the intertidal zone. However, the watersheds to the south of the coastal zone are managed by a USAID programme that aims to stabilize the watershed slopes, reduce erosion, and promote sustainable agriculture, so effectively the area is under a sustainable management regime from ridge to reef.

The PN3B is Haiti's first marine protected area. In addition to being a key biodiversity area, the bays of Limonade, Caracol and Fort Liberté in Haiti, together with the Monte Cristi area in the neighbouring Dominican Republic, form one of 22 Ecologically or Biologically Significant Marine Areas (EBSAs) that have been identified throughout the wider Caribbean and western mid-Atlantic. The protected area is an important component of the Caribbean Biological Corridor that runs from Cuba to the Dominican Republic under the protocol concerning specially protected areas and wildlife (SPAW).

The mangroves, turtle and manatee grassbeds, reefs and bay habitats are important nurseries for economically important fish, crustaceans and mollusks. This area is crucial for subsistence fisheries for local communities. The reefs serve to protect the low-lying plains from storm surges, and the mangroves provide protection from wind and waves. In particular, the mangrove forests of Caracol and Fort Liberté Bays play an important role in the reproductive cycles of numerous coastal and pelagic fish species, including those important for human consumption such as snook, prawns and lobsters and key mollusks such as the queen conch. At least 13 species of reptiles and birds considered either threatened or seriously in danger of extinction have been identified as potential inhabitants of mangrove forests and lagoons of the area. Among those are the American crocodile, the West Indian manatee, the Atlantic leatherback sea turtle, the Atlantic Hawksbill sea turtle; and amongst birds, the flamingo, the black-crowned palm tanager, the northern mockingbird and the cave swallow. Of these, the black-crowned palm tanager is endemic to Hispaniola and the cave swallow is endemic to Haiti.

Apart from its rich biodiversity, the PN3B has many cultural and historical sites of interest. Limonade is close to the ruins of the sixteenth century Fort Puerto Real; Fort Liberté (formerly Fort Dauphin) is next to the old Spanish Fort Bahaya; and the Caracol area (Bas Saline), according to many archeologists, is the location of Fort La Navidad, the first Spanish settlement founded by Christopher Columbus in the New World in December 1492.

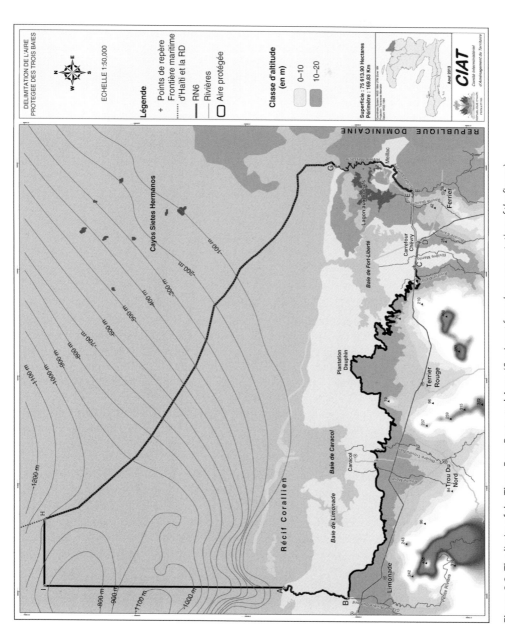

Figure 3.2 The limits of the Three Bays Protected Area. (*See insert for color representation of the figure.*)

Coastal Threats to the PN3B

Threats to coastal and marine environments and their biodiversity are generated firstly by the coastal communities that depend on the ecosystems for their livelihoods, businesses and recreation, and secondly by the impact of climate change: rising sea-levels, higher sea surface temperatures and ocean acidification. The threats to the coastal and marine ecosystems within the PN3B come from:

- Overfishing
- Destruction of mangroves
- Eroding watersheds
- Nutrification
- Coastal construction
- Pollution
- Invasive species
- Disease
- Sea-level rise
- Increasing sea surface temperature
- Ocean acidification.

Overfishing

The primary threat to the marine environment in Haiti is overfishing. In 1926, the New York Zoological Society sent a team of scientists to Haiti to study coral reef fish using helmet diving (Beebe, 1928). At that time, there were abundant large fish of all species from large parrotfish to big sharks and 2 m long goliath grouper and tarpon on reefs just west of Port-au-Prince. Ichthyologist William Beebe wrote, 'A mob of huge parrotfish came into view ... 300 ... most were more than two feet and at least 20 measured a full yard.'

But by 1986, in a short paper published entitled 'Overfishing: the Haiti experience' the authors noted that large fish were never observed on or above the reefs of La Gonave, even though at that time, according to the authors, the reefs themselves were still in good condition.

The situation now is that every shallow coral reef in Haiti is being severely overfished, and that overfishing has destabilized most reefs to the point where they have been turned into algae-dominated rocky cemeteries. Overfishing down the food chain, from predators to herbivores, eventually reduces the abundance of algae-eating fish that keep the reef in good health. Without sufficient fish herbivores, the algae keep growing and simply spread over the reef corals, smothering most of the living hard coral leaving them like rocks covered in algae. Most of the coral reefs of Haiti are now in this degraded condition: algal-covered reefs with almost no fish longer than 20 cm in length.

While overfishing also occurs in seagrass beds and mangrove forests, the major impact is on coral reefs where fish do not have a chance to reach reproductive maturity and so do not reproduce before they are caught – a vicious circle of overfishing that inevitably leads to fewer and smaller fish. It is an unfortunate characteristic that many reef fish sought for food such as grouper need to live up to six years and attain a size of 50 cm before they become reproductively mature.

In the Three Bays area, overfishing is worse between Limonade and Caracol due to the greater number of fishermen working in that zone. More than 100 fishing boats go out through the reef pass each morning. This is reflected in the condition of the barrier reef and patch reefs in the Limonade and Caracol Bays, which are algal dominated. The reefs at Fort Liberté are also overfished but are in better condition with a higher percentage of living coral cover. The fishing pressure is less due to a smaller number of fishers.

Destruction of Mangroves

The Caracol Bay area includes an estimated 3900 ha of healthy mangroves (approximately 18% of the remaining mangrove in Haiti), turtle and eel grass beds, and a sheltered bay protected by a coral reef that extends over 20 km.

Almost everywhere in Haiti there is widespread exploitation of mangroves for firewood, charcoal and building materials. The destruction of mangrove forests has a huge impact on the resilience of the coastal environment and its ecosystems. In addition, the establishment of evaporation ponds for the production of sea salt has caused further destruction of the mangrove areas around Caracol and Fort Liberté. Figure 3.3 shows mangroves being cut for charcoal in the area of the Three Bays Park.

Eroding Watersheds

Haiti is well known for its deforestation and eroded watersheds. When it rains, sediment will eventually wash into the sea either directly or via streams and rivers. But the deteriorating situation on land is not replicated in the sea. Based on surveys conducted by the organization Reef Check, it appears that only about 10% of Haiti's coast is badly affected by sedimentation. In these areas, so much sediment is being discharged or resuspended in the water column by waves that the corals suffer due to the lack of light, sediment deposition that prevents larval settlement and attachment, and the inability of corals to clean their surfaces as fast as sediment is deposited. When sediment builds up on the coral, bacterial infections start and eventually lead to the coral's death.

Most corals can survive moderate levels of chronic sedimentation. It is the acute exposure due, for instance, to a hurricane that washes large amounts of sediment into the sea that can quickly kill a coral, especially if it is buried in sediment. However, a moderate amount of sedimentation is beneficial for both seagrass beds and mangrove forests.

Eroded watersheds cause another serious problem. Flooding of the coastal towns is frequent. Limonade is close to the Grande Rivière du Nord, a river that discharges into the Limonade bay and which regularly floods the town. The town of Trou de Nord is frequently flooded by the river of the same name, and the town of Caracol is regularly flooded during periods of heavy rain. The flooding of the coastal towns washes huge quantities of polluted water into the ocean, conveying tons of plastic trash and garbage. If the flooding of the coastal urban areas was eliminated, the management of sewage, household pollution and plastic trash could be more effective and the impact on coastal ecosystems much reduced.

Figure 3.3 Charcoal production from mangroves close to the PN3B. *Source:* Courtesy of Andy Drumm, Drumm Consulting. (*See insert for color representation of the figure.*)

Nutrification

Normally, coral reef waters are oligotrophic with few nutrients. Small amounts of nutrient added to the system are quickly used by algae and seagrasses. Excess nutrients can build up in the water column due to runoff of agricultural fertilizers and/or sewage from humans or other animals. Excess nutrients will stimulate plant growth and this tends to favour the growth of algae (and sponges) to the detriment of corals. Nutrification often goes hand-in-hand with sedimentation because the sediment carries additional nutrients.

Until 2012, when the first sewage treatment plant was built outside of Port-au-Prince, there was no sewage treatment in Haiti. So essentially all human waste ended up in drainage channels that empty into the sea during heavy rainfall.

In the PN3B, in areas near Fort Liberté and close to the mangroves, the water can become green, eutrophic and turbid when freshwater inputs are high and when wind resuspends sediment. A similar phenomenon occurs in the Limonade and Caracol Bays where there are far more mangrove and freshwater inputs. Lagoon Aux Boeufs, which is almost entirely enclosed, appears to be in a constant state of eutrophication based on satellite images.

Coastal Construction

The Caracol Industrial Park will generate employment in the region, and the coastal strip between Limonade and Fort Liberté is attracting Haitians looking for work. Investments in tourism infrastructure and services planned for the area around Fort Liberté will also attract men and women seeking employment. Housing for low-income families in Haiti is generally poor quality and lacks basic sanitation. Many homes do not have toilets. Sewage and fecal material finishes up in local rivers and is washed out to sea. Much of the coastal strip is private land, so new housing construction may be on land cleared of mangroves, or on mudflats north of the main coastal road. New hotels will spring up along the coast. In the absence of building codes enforced by local authorities, most of these hotels will lack basic sanitation infrastructure.

In some locations, seagrass beds, mangroves and coral reefs may be directly destroyed by the construction of roads and buildings, the cutting and infilling of mangrove forest and seagrass beds, and dredging of coral reefs to create channels. Even when great care is taken, there are some areas where some corals or associated habits will be damaged either directly or indirectly by sediment transport and deposition.

Pollution

In addition to sedimentation, sewage and nutrients, there are several other types of pollution that can affect the marine environment. The marine systems adjacent to towns such as Fort Liberté are exposed to all types of pollution from fuel oil to solid waste. Many chemicals are toxic to seagrass beds, mangroves and coral reef organisms. Solid waste can be a problem for sea turtles and sea birds which can become entangled, or they eat small coloured bits of plastic until their digestive systems are effectively destroyed. Tons of floating plastic waste are washed out to sea from coastal towns in Haiti every time there is heavy rainfall along the coasts (see Figure 3.4).

Invasive Species

A number of invasive marine species have come into the Caribbean from other areas, but perhaps none so damaging as the lionfish (*Pterois volitans*). This scorpion fish from the IndoPacific is a popular 'pet' for home aquarium owners and was probably released into the sea in Florida accidentally. Unfortunately, the lionfish has no predators and has extremely poisonous spines. It is a voracious predator with a large mouth, and consumes juvenile fish of many species. Large lionfish are now common on most reefs in Haiti.[2]

---

2 The good news is that lionfish are on the menu in the Dominican Republic (DR) and the practice has spread to Haiti. There is even a lionfish festival in the DR once a year in January. This is by far the best way to reduce the numbers of an invasive marine species. It also provides an income for fishers whose livelihoods may be curtailed by restrictions on fishing in some marine park areas in order to protect the reproduction and development of fish species essential for a healthy marine ecosystem.

Figure 3.4 Plastic bottles and floating garbage washed up on one of the small islands on the west coast of Haiti. *Source:* M. Bush. (*See insert for color representation of the figure.*)

Diseases

During the late 1970s and early 1980s, two different diseases spread throughout the Caribbean and killed off the long-spined black sea urchin and two species of coral, the staghorn and elkhorn. These two species of coral were so abundant that they formed two distinct extensive zones on the reef. When these two coral ecosystems died out, the reefs were missing two broad bands of coral accounting for 30–40% of the area. The loss of the black sea urchin further exacerbated the overfishing problem with respect to herbivores, and resulted in more macroalgae taking over the reefs. These diseases did not affect seagrass beds or mangrove forests. At this time, it appears that the diseases have stopped killing the corals and sea urchins, and both are making a slow come-back. The fact that there are significant numbers of colonies of staghorn and elkhorn corals in Haiti is extremely important because the reefs can serve as a source of larvae for both corals and urchins in downstream areas. The faster the recovery of the reefs in Haiti, the faster the recovery will be in Florida.

Sea-level Rise

It is certain that climate change will cause sea-levels to rise throughout the Caribbean. There is evidence that coral reefs can adapt naturally to this slow rise *if* they are not overstressed by anthropogenic factors such as overfishing, pollution, extraction and excessive sedimentation. On land, the main threat is that the natural tendency of mangroves to migrate inland to a slightly higher elevation to compensate for the higher sea-level is blocked by hard construction and infrastructure right behind and too close to the edge of the mangroves.

Increasing Sea Surface Temperatures

Little can be done locally to reduce the impact of higher ocean temperatures on coral reefs except to try to lessen the levels of additional stresses such as pollution, excessive sedimentation and overfishing. But nature is inherently resilient, and if these land-based stressors can be eliminated, there is a good chance that coral reefs can survive.

Mapping the Threats

Ecosystem-based adaptation to climate change impacts on coastal environments starts by identifying and mapping all the threats to marine and coastal ecosystems. These threats come from the ocean, from low elevation coastal zones, and from the watersheds and catchment areas that channel water to the coast and out to sea. In the case of the Three Bays Park, several of the threats identified above originate in the watersheds south of the coastal zone, particularly the large quantities of sediment from eroding watersheds conveyed to the coast and out to sea during heavy rainfall. The ridge-to-reef (R2R) concept provides a planning tool essential for EBA planning. The R2R methodology does not mean that an EBA programme will necessarily plan on implementing management action in the whole of the area defined by the R2R approach. But the R2R analysis provides the basis and justification for the terrestrial actions judged to be a priority in order to restore and build resilience in coastal and marine ecosystems.

All the watersheds contiguous with the coastal zone should be identified and mapped from ridge to reef. For the coastal zone associated with the PN3B there are seven watersheds. Figure 3.5 outlines the limits of the watersheds and shows the main towns in the area, each of which is the seat of one of the nine Communes that are located in the area.[3]

Seven watershed areas are shown on the map in Figure 3.5:

1) The watershed of the Grand Rivière du Nord, which discharges on the western edge of the Bay of Limonade.
2) A catchment area to the east of Limonade that includes several small rivers and streams that discharge into a broad wetlands area to the west of Caracol.
3) A relatively small watershed discharging into the Bay of Caracol to the west of the town.

---

3 A Commune in Haiti is equivalent to a municipality in the UK or a County in the US. A Commune has an elected mayor, and is the level of local government capable of mobilizing local communities and community-based organizations collectively to tackle climate change issues.

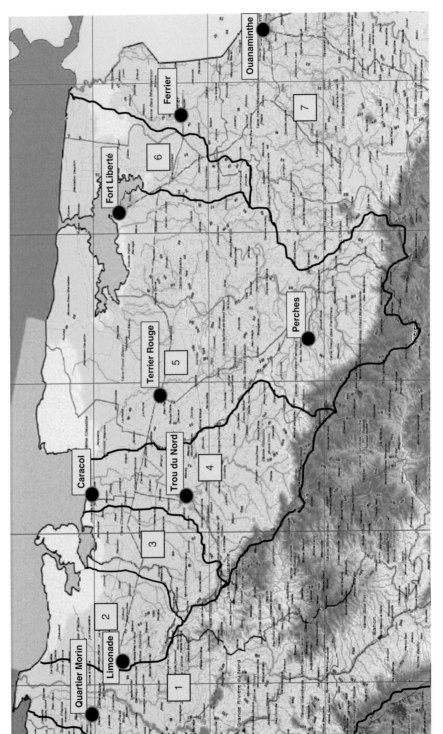

Figure 3.5 The seven watersheds linked to the Three Bays Protected Area (PN3B) in Haiti. (*See insert for color representation of the figure.*)

4) The major watershed of the Trou du Nord river, which discharges close to Caracol.
5) The watershed of the River Marion, which discharges to the west of Fort Liberté. Several smaller rivers also run through the watershed discharging into the Fort Liberté Bay further to the west.
6) The most eastern of the watersheds that discharges into the sea north of the plain.
7) A complex of watersheds which run northeast towards Ouanaminthe and which all discharge into the Massacre River, which marks the border with the Dominican Republic.

The six Communes that are situated within the ridge-to-reef zone are shown in Table 3.5.

Given the employment opportunities in the area, the low population density in the Caracol Commune is striking, perhaps pointing to large areas of private land on the coastal plain which are off-limits to local farmers.

The threats to the coastal and marine areas are mapped onto a topographic map of the zone as shown in Figures 3.6 and 3.7. The origin of the threat is referred to as the impact generating area or IGA. The area impacted is called the impact receiving area or IRA. IGAs and IRAs are connected by impact flow areas through which the potential impact is transmitted. Areas that may actually benefit from the threat are identified as benefit receiving areas or BRAs (Bush, 2014).

The management of the threats to coastal ecosystems can then be spatially disaggregated and assigned to a governance centre: in this case the Communes located in the six watersheds that discharge directly into sea on the northeast coast of Haiti.

The majority of the threats confronting the coastal and marine ecosystems along the northeast coast of Haiti are manageable; they can be reduced in several ways. Table 3.6 outlines the steps that need to be taken.

The design and implementation of ecosystem-based adaptation management plans requires the full collaboration of all the stakeholders including local communities and community-based organizations. This is a huge challenge, but there are many examples among the SIDS where this challenge has been met and overcome.

The Three Bays Protected Area in Haiti is just the first of several marine protected areas planned in that country over the last few years. But protected areas legally established on paper and then never provided with the financial resources and management

Table 3.5 The communes defining the ridge to reef zone.

Commune	Population	Area, km^2	Population /km^2
Fort Liberte	32,861	240.3	137
Ferrier	13,973	70.0	200
Trou du Nord	46,695	130.8	357
Terrier Rouge	28,938	171.2	169
Caracol	7362	74.9	98
Perches	11,028	39.6	278
Totals	140,857	726.8	194

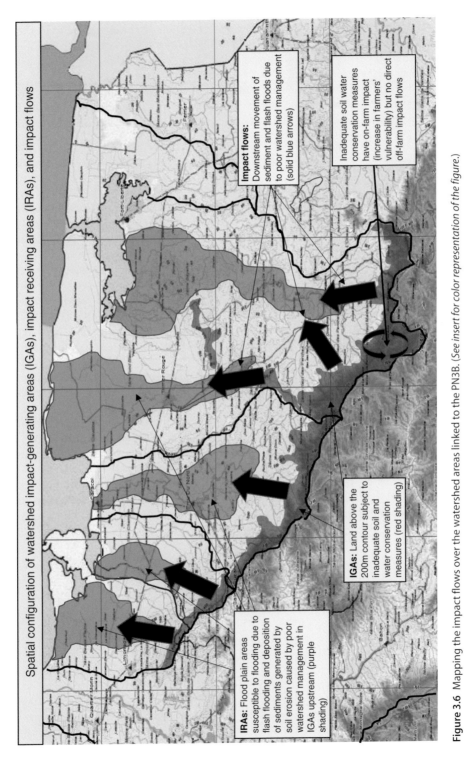

Figure 3.6 Mapping the impact flows over the watershed areas linked to the PN3B. (*See insert for color representation of the figure.*)

Spatial configuration of watershed impact-generating areas (IGAs), impact receiving areas (IRAs), and impact flows

Impact flows:
Downstream movement of sediment and flash floods due to poor watershed management (solid blue arrows)

Inadequate soil water conservation measures have on-farm impact (increase in farmers' vulnerability) but no direct off-farm impact flows

IGAs: Land above the 200m contour subject to inadequate soil and water conservation measures (red shading)

IRAs: Flood plain areas susceptible to flooding due to flash flooding and deposition of sediments generated by soil erosion caused by poor watershed management in IGAs upstream (purple shading)

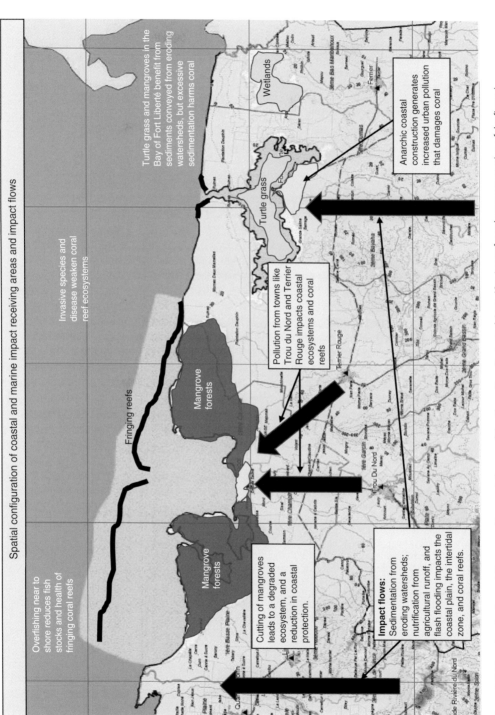

Figure 3.7 Mapping the impact flows over the coastal areas of the Three Bays Protected Area. (*See insert for color representation of the figure.*)

Table 3.6 Managing the threats to the Three Bays Protected Area.

Overfishing

The marine environment needs to be managed in the same way as the terrestrial environment. Areas need to be zoned and their use restricted; some areas should be off-limits to fishers. Defining zonal arrangement, and respecting the regulations that apply, requires the full participation of the local communities and the associations of fishers. A good example of where this community-based co-management has worked is the La Caleta marine park in the Dominican Republic where fishers are now engaged as tour guides, diving instructors and boat operators for tourists.

Disease

Many diseases that affect reef ecosystems are caused by a combination of anthropogenic stress, which can be reduced with good environmental management, and by climate change stress factors that, being at a larger scale and longer term, are more difficult to mitigate. Management objectives should focus on reducing stress caused by pollution, eutrophication and excessive sedimentation, while at the national government level, efforts should continue to reduce emissions of greenhouse gases.

Destruction of mangroves

Coastal communities need to better understand the importance of mangrove forests. Cutting mangroves should be prohibited throughout Haiti, and the Government should be promoting public awareness campaigns that explain the importance of these trees. At the local level, all mangrove forests should be in coastal and marine managed areas that are co-managed with local communities.

Eroding watersheds

Each of the six watersheds needs to have a management plan that aims to reduce erosion and flooding by increasing the areas that are under sustainable management, including areas left to grow back naturally and which are never again cleared for subsistence agriculture. The management of watersheds must be conducted with the full participation of local communities and community-based organizations.

Nutrification

Reducing nutrification depends on the widespread adoption of agricultural best practice. The use of fertilizer is being increasingly promoted in Haiti by the Ministry of Agriculture (MARNDR). The Ministry has local agents in the Communes that should work with farmers to ensure that fertilizers are used correctly and that the quantities applied are not excessive.

Coastal construction

The Communes are responsible for issuing construction permits and ensuring that new houses have basic sanitation systems. The Communes should shoulder this responsibility and enforce the regulations. For major tourist infrastructure, environmental impact assessments (EIAs) are essential and the ministries responsible (Environment and Public Works) should work together to ensure that the construction of hotels, restaurants and other infrastructure complies with the appropriate regulations. EIAs are particularly important when proposed infrastructure will directly impact the coastal zone, such as jetties, wharfs, marinas and port infrastructure.

Pollution

The amount of trash washed out to sea around Haiti is huge. This is a major problem that needs to be tackled at national government level, working in collaboration with the private sector invested in beach tourism.

Invasive species

This problem is difficult to manage. What is important is to identify the problem early and to evaluate and implement remedial action based on scientific evidence. If the species in question is edible (like the lionfish), the government working with the private sector should organize high-profile media events where the fish are cooked and served.

expertise to enable them to be properly and sustainably managed are the norm in developing countries, rather than the exception, and Haiti is no different. These paper parks are not really parks at all, and very few of them are adequately managed or protected.

The protection and sustainable management of coastal and marine areas should be an absolute priority for small island governments. The full extent of the coastal and marine environment around small island states needs to be legally protected and managed according to a published management plan. The coastal and intertidal areas around every island need to be protected and sustainably managed. The present piecemeal approach of establishing individual marine protected areas in sites considered to be biodiversity hotspots is only the starting point for a programme that brings the *entire coastline* and its marine contiguous zone under a much larger national plan that identifies the different coastal zones and the regulations that apply within them. For states that include dozens of islands, it may not be feasible to include them all in a management plan, but zoning all the main islands should be an immediate priority.

The Management of Coral Reefs

To help preserve the diversity of the coral reef ecosystem, a protected reef area should contain many different habitats for a steady and varied supply of larvae to replenish naturally damaged areas and to replace dead or emigrated organisms. This is important to help maintain a source of larvae to reestablish parts of the reef destroyed by bleaching events. In practice, three kinds of habitats should be included in coral reef reserves: coral habitats, neighbouring coastal habitats (i.e. submerged, intertidal, or above water), and distant habitats that are linked to the reef (Salm *et al.*, 2000).

Neighbouring habitats should include:

Reef flats	• Which introduce fixed nitrogen, dissolved and particulate organic compounds into the reef food web; • function as feeding grounds and nurseries for reef fishes, increases diversity and abundance of reef species.
Seagrass beds	• Which introduce dissolved and particulate organic compounds into the reef food web; • provide feeding grounds and nurseries for reef organisms to increase diversity and abundance of reef species; • consolidate sediments, protecting the reef from being smothered.
Sand or mud flats	• Provide feeding grounds for reef fish, increasing the diversity and abundance of reef species.
Mangroves, lagoons and estuaries	• Which introduce dissolved and particulate organic compounds into the reef food web; • increase diversity and abundance of reef species; • trap pollutants and silt, protecting the reef from being poisoned or smothered.

In principle, the optimal size of a protected reef area is a strictly controlled core encompassing a sufficient area of reef to be self-replenishing for all species. This is particularly important if preserving biological diversity is the principal management objective. This design criterion is less important for other objectives: for instance, maintaining the area's value for recreation or tourism, or as an area that is protected and managed so as to block overfishing and allow fish stocks to recover. The core zones of coral reef protected areas should not be less than about 450 hectares (Salm *et al.*, 2000), but protected coral reefs should always be part of a much larger marine protected area that includes the coastal zone and as much of the coastal watersheds as possible.

For the smaller islands, *the whole island including the entire coastal zone, the island interior and all the watersheds should be covered by a management regime.*

Where islands are larger, the coastal zone under management is a broad belt of protection that should still circumscribe practically the whole island. The majority of coastal zones would be protected landscapes where sustainable use of natural resources is permitted. Industrial and commercial areas – ports, marinas, harbours and the like – are designated as such but still controlled so as to prevent pollution and the unauthorized discharge of liquid and solid waste. Beach and tourist areas are marked out and controlled. Strictly no-use areas around vulnerable coral reefs and threatened marine ecosystems are identified and more carefully managed.

There is evidence that, where coral reefs are sustainably managed by local communities that have customary rights over the marine resources, the reefs are more resilient and adapt more effectively to the stress of increased water temperatures (Cinner *et al.*, 2016). This finding supports the assertion that coral reefs are much better protected if coastal pollution is eliminated, overfishing curtailed, and the area sustainably managed.

Climate Change Adaptation in Agriculture

The production of food crops is globally one of the most important economic activities, and also one of the most sensitive and vulnerable to climate change. Agriculture in developing countries is already being negatively affected by climate change, and the situation is likely to worsen given the latest forecasts. Global temperatures will increase, patterns of precipitation will become unpredictable, and pests and diseases will becomes more difficult to manage. Impacts will not be uniform across the regions of the SIDS, and different islands will require different adaptation strategies depending on their situation.

Climate-smart Agriculture

In principle, increased levels of atmospheric CO_2 should enhance photosynthesis efficiency and reduce respiration rates, offsetting the loss of productivity due to higher mean temperatures. However, field-scale experiments have not corroborated this hypothesis (Long *et al.*, 2006) In addition, by the time CO_2 levels have risen to the point where they will potentially make an impact on productivity, local temperatures will also have risen to a level where any gain in productivity is likely to be cancelled out (Turral *et al.*, 2011).

The effect of temperature changes on agricultural yields and biomass production levels is complex. Increasing temperatures that lead to a warmer soil at planting can contribute to quicker germination and growth, increasing biomass production; and a longer and warmer growing period may extend the growing season and the potential for increased biomass production. However, if the temperature is too high and is accompanied by drier periods and droughts, the resulting stress can result in lower biomass production (Delgado *et al.*, 2013).

Impacts will not be uniform across the regions of the SIDS, and different islands will require different adaptation strategies, but in general the poorer islands will face the greatest challenge in trying to prevent present levels of food insecurity from becoming more severe.

Maintaining and increasing food security requires agricultural production systems to become more productive, more resilient, and capable of performing well in the face of disruptive climate events. This transition requires transformations in the management of natural resources – land, water, soil nutrients, and genetic resources – and greater efficiency in their use and the inputs for production. There are mitigation benefits also; climate-smart agriculture will generally increase carbon sinks as well as reducing emissions per unit of production.

Climate-smart practice is needed in both commercial and subsistence agriculture, but there are significant differences in the approach. In commercial systems, increasing efficiency and reducing emissions are the main concern. In poorer countries where subsistence agriculture is the principal livelihood for the majority of the population, transforming smallholder systems is important for maintaining and enhancing food security and reducing poverty (CGIAR, 2014).

The FAO has defined climate-smart agriculture as agriculture that sustainably increases productivity and resilience (adaptation), reduces/removes greenhouse gases (mitigation), and enhances the achievement of national food security and development goals (FAO, 2010a). Climate-smart agriculture (CSA) thus encompasses both adaptation and mitigation.

The efficiency, resilience, adaptive capacity and mitigation potential of agricultural production systems can be strengthened by improving the effectiveness of its different components: the key ones are outlined below.

Soil and Nutrient Management
The availability of nitrogen and other nutrients is essential to increase yields. Techniques include: composting crop residues and manure; improved matching of nutrients with plant requirements; controlled release and deep placement techniques; and using legumes for nitrogen fixation. Practices that increase organic nutrient inputs reduce the need for synthetic fertilizers – a reduction that comes with several co-benefits. Subsistence agriculture is often practiced on poor soil that have low nutrient content. Soil quality can be improved by intercropping legumes as green manure, crop rotation using legumes, and agroforestry systems incorporating nitrogen-fixing trees.

Water Management
Improved water collection and use of pools, dams, pits, bunds and ridges, and increasing the efficiency of water use, are essential for maintaining levels of production in the face of increasingly variable and unpredictable patterns of rainfall. Irrigation is practiced on only about 20% of arable land in developing countries, but irrigated areas can produce over twice as much as rainfed fields (FAO, 2010b). Good water management need not require expensive infrastructure; in the Yatenga province of Burkina Faso,

farmers reclaimed degraded farmland by digging planting pits called *zai*. This traditional water management technique was improved by increasing the depth and the diameter of the *zai* pits and mixing in organic matter. Land that was almost barren can now achieve yields of up to 1500 kg/ha depending on rainfall. Stone contour bunds have also been built to harvest rainwater. The bunds spread rainwater across the fields, increase infiltration, and reduce the loss of organic matter (Reij *et al.*, 2009).

Pest and Disease Control

There is evidence that climate change is altering the distribution, incidence and severity of animal and plant pests and diseases, as well as the occurrence of invasive and alien species. The development of disease- and pest-resistant varieties will become increasingly important.

Resilient Ecosystems

Improving the management of ecosystems and biodiversity provides a wealth of benefits, and evidence is growing that restoring and strengthening local ecosystems must be a key element of climate-smart agriculture.

Genetic Resources

Plant genes determine a plant's resistance to shocks such as temperature extremes, drought, flooding, pests and diseases. They also regulate the length of the growing season and the plant's response to fertilizers and water. The preservation of genetic material is therefore essential in developing plant varieties that are more resilient and which use inputs more efficiently. For example, the most common adaptation strategies used by farmers in Ethiopia and South Africa have included introducing new crop varieties (Bryan *et al.*, 2009). Where salinization is a problem due to seawater incursion, the introduction of salt-tolerant crops and cultivars helps to ensure that agricultural production is more resilient. In Haiti, the FAO has introduced a bean variety called ICTA Ligero, which matures early and is resistant to the Golden Mosaic virus – prevalent in Haiti. In 2009, the FAO seed multiplication programme supported 34 seed producer groups that produced 400 tonnes of bean seed, including ICTA Ligero (FAO, 2010a).

Conservation Agriculture

Conservation agriculture (CA) refers to the management of agriculture and agro-eco-systems to increase sustainability and productivity, and to increase food security, while restoring and strengthening the resource base and the environment.

Conservation agriculture has three key characteristics:

1) minimal disturbance of the soil by ploughing or other means;
2) maintaining soil cover with crop residues or cover crops;
3) diverse cropping systems, such as rotation, multicropping, intercropping and agroforestry.

Ploughing the soil or tillage has been a standard practice in agriculture for centuries. Although it increases productivity in the short term, over the longer term it leads to a reduction of soil organic matter, and most soils degrade under prolonged intensive arable agriculture. This degradation results in the formation of crusts and compaction, and leads finally to soil erosion. The mechanization of soil tillage – which allows greater working depths and the use of ploughs, disc harrows and rotary cultivators – has had particularly detrimental effects on soil structure. Movements to reduce tillage have gained momentum over the last two decades and the technologies have been adapted for nearly all soils, crop types and climatic zones (FAO, 2015).

In soil that is not tilled for several years, crop residues remain on the surface of the soil and produce a layer of mulch. This cover protects the soil from the physical impact of rain and wind, reduces the evaporation of soil moisture, and stabilizes moisture and temperature in the surface layers. The mulch and topsoil becomes a habitat for several organisms including insects, fungi and beneficial bacteria. These organisms break down and decompose the mulch, and mix it in with the soil to form humus, which contributes to the physical stabilization of the soil structure and provides a buffer for water and nutrients. Soil fauna such as earthworms help to create stable and porous soil that allows faster water infiltration under heavy rain.

The advantages of soil cover include:

- Improved infiltration and retention of soil moisture, resulting in less severe and less prolonged crop water stress and increased availability of plant nutrients.
- Providing a source of food and habitat for diverse soil organisms by creating channels for air and water, and by recycling organic matter and plant nutrients.
- Increased humus formation.
- Reduction of impact of raindrops on the soil surface, resulting in less surface crusting.
- A reduction in runoff and erosion.
- Soil regeneration is greater than soil degradation.
- Temperature variations in the soil are reduced.
- Better development of roots and seedling growth.

Keeping the soil covered is a basic principle of conservation agriculture. Crop residues are left on the surface, but cover crops may need to be planted if the gap is too long between harvesting one crop and planting the next. Cover crops are grown mainly for their effect on soil fertility or as livestock fodder. Cover crops will:

- Protect the soil when it does not have a crop.
- Provide an additional source of organic matter that improves soil structure.
- Recycle nutrients and make them available to subsequent crops.
- Provide 'biological tillage' where the roots of some crops are able to penetrate compacted soil and improve water infiltration.
- Reduce the leaching of nutrients such as nitrogen (FAO, 2015).

Biomass material left in contact with the soil has a much greater impact than canopy cover because it not only protects the soil from raindrop splash, but also substantially increases surface water flow roughness, which reduces both flow velocities and the ability of water to move sediment.

It is important to start the first years of conservation agriculture with cover crops that leave substantial residues on the soil surface, and which decompose slowly. Grasses and cereals are the most appropriate plants for this phase because of their aggressive and abundant rooting system. In the following years, legume cover crops can be added. These enrich the soil with nitrogen and decompose more rapidly. Later still, when the soil is stabilized, cover crops with an economic value such as fodder for livestock can be incorporated into the rotation system (FAO, 2015).

No-till agriculture requires direct seeding, which involves growing crops without mechanical seedbed preparation and with minimal soil disturbance since the harvest of the previous crop. There is mechanical equipment available to carry out this form of planting. It can also be done by hand.

In Haiti, no-till agriculture is a relatively new conservation agriculture approach. The SANREM[4] programme run by Virginia Tech has been experimenting with the technique for several years, and has had good preliminary results, as shown below for experimental plots in Corporant and Lachateau. In Corporant, maize yields were slightly less the first year, but increased in subsequent seasons.

The conclusion was that soil conservation practices are positively correlated with crop yields. They also found that reduced tillage does not reduce crop yields; that cover crops have little impact on crop yields; and that weed populations are lower with a cover crop and no ploughing. No-tillage approaches to conservation agriculture have been shown to increase the retention of moisture in the soil. Research by Virginia Tech in the Philippines showed that water retention increased significantly with no-tillage agriculture.

The benefits of conservation agriculture can be grouped into three categories:

- Economic benefits that improve production efficiency.
- Agronomic benefits that improve soil productivity.
- Environmental and social benefits that protect the soil and make agriculture more sustainable.

Economic Benefits

These include a reduction in labour requirements; a reduction in costs, for example in operating machinery and reduced labour costs; and high efficiency in terms of a greater output for a lower input.

The positive impact of conservation agriculture on the distribution of labour during the production cycle, and a reduction in labour requirements, are the main reasons why farmers in Latin America have adopted CA. Back-breaking manual labour for soil preparation is unnecessary (FAO, 2015).

Agronomic Benefits

Soil productivity is improved due to the increase in organic matter, better soil-water conservation, and improved soil structure. The constant addition of crop residues leads to an increase in the organic matter content of the soil. At first this is limited to the top layer of the soil, but over time this improvement extends into deeper soil. Organic matter improves fertilizer use, water-holding capacity, soil aggregation, rooting environment and nutrient retention.

---

4 SANREM stands for sustainable agriculture and natural resources management. SANREM is funded by USAID.

Environmental and Social Benefits

Soil erosion is greatly reduced with conservation agriculture, resulting in less down-stream impact on coastal zones and water retention infrastructure including hydroelectric installations. Groundwater quality improves, biodiversity increases, and carbon sequestration can increase markedly.

Crop residues remaining on the surface absorb the splash effect of raindrops, which then infiltrate to the underlying soil with much less destructive impact. The biomass residues form a physical barrier that slows runoff and reduces the evaporation of soil moisture. Soil erosion is substantially reduced. More water infiltrates into the soil than runs off the surface, which increases subsurface flow. Streams have greater volumetric flow and the water is cleaner. Higher rates of infiltration reduces flooding, recharges groundwater and feeds springs.

The most direct effect of climate change on rainfall-driven erosion is related to the erosive power of rainfall. However, the dominant variable appears to be rainfall intensity and energy rather than rainfall amount alone (Nearing *et al.*, 2005). This result suggests that extreme rainfall events (storms and cyclones), generally projected to increase by climate change models, will have a much greater effect on soil erosion than any regional trend towards higher rates of annual precipitation.

Crop Rotation

A diversity of crops in rotation leads to a diversity of soil flora and fauna, as the roots excrete different organic substances that attract different types of bacteria and fungi which, in turn, play an important role in the transformation of these substances into plant nutrients. Different crops root at different levels, allowing nutrients at deeper levels to be recycled to the surface. The benefits of crop rotation include:

- Higher diversity in plant production and thus in human and livestock nutrition.
- Reduced risk of pest and weed infestations.
- Greater distribution of channels or biopores created by diverse roots.
- Better distribution and utilization of water and nutrients through the soil.
- Increased nitrogen fixation through certain plant–soil interactions and improved balance of N/P/K from organic and mineral sources.
- Increased humus formation (FAO, 2015).

Landscape Impacts

An important aspect of conservation agriculture is its ability to change the landscape dramatically. The removal or destruction of a vegetative cover has a negative impact on local plants and animals. Conservation agriculture maintains a habitat for species that feed on pests, which in turn attracts more insects, birds and other animals. Biodiversity is strengthened and made more resilient.

Cropping systems based on maintaining crop residues and no tillage will accumulate higher levels of carbon in the soil, compared with the emissions to the atmosphere resulting from plough-based tillage. During the first years of implementing CA, the organic matter of the soil is increased through the decomposition of roots and vegetative cover on the surface. The organic matter decomposes slowly, but much of it is incorporated into the soil profile, so the emission of carbon is also slow. The result is that carbon is sequestered in the soil which becomes a net sink of carbon. The introduction and practice of conservation agriculture is therefore not only an important form of adaptation to climate change, but holds the promise of substantially reducing emissions of CO_2 from agriculture. It has been calculated that the total potential for carbon sequestration by conservation agriculture could offset about 40% of the estimated annual increase in CO_2 emissions (Robbins, 2004).

Soil Erosion

While no-till agriculture clearly reduces soil erosion, in many developing countries more traditional practices have been applied for many years. For example, in Haiti, where agriculture is frequently on steeply sloped land, barrier methods of soil conservation have been used for decades. The most common is a rock wall or dry wall terraces which are built along a contour and which are effective if properly constructed and maintained. This is a common practice where rocks are easily available on the surface, but they require a significant amount of labour and time. On many mountain slopes rocks are abundant and so the practice is common. A variation, known as *cordons de pierre*, is simply a line of rocks running along a contour. In basalt areas where rocks are not so easily available, contour hedgerows are a better option.

Deep contour canals have been promoted but are not widely used. Contour hedgerows are often planted with *Leucaena leucocephala*, which is nitrogen-fixing. Hedgerows can be effective if well managed, and bench terraces develop rapidly behind them.

Banje manje are contour plantings of semi-perennial and long-term food crops such as bananas, plantains, pineapple, and sometimes malanga and cassava. There is also a practice of creating barriers with crop residues partially incorporated into the soil. Rows of forage bunch grasses have also worked, including vetiver.

The pie chart in Figure 3.8 shows the relative frequency of conservation practices surveyed on the central plateau of Haiti (Thompson *et al.*, 2014).

The baseline surveys conducted by Virginia Tech in Haiti also found that:

- plots are highly intercropped, the most common crops being corn, sorghum, pigeon pea, manioc, banana, squash, peanut and okra;
- about 40% of plots are prepared for planting with an ox-driven plough; 1% of farmers used a tractor, while the remainder used hand tools;
- about 40% of households had at least one soil conservation practice on one or more of their plots;
- households are more likely to establish 'live' barriers on plots they perceive as having poorer soil, and they are more likely to establish 'dead' conservation practices (e.g. rock walls or barriers) on plots they perceive as having better soil;
- land tenure status does not appear to be a significant incentive or deterrent to the adoption and use of common soil conservation practices.

The survey also found that agricultural incomes increased by 79% with rock walls and 39% with the use of hedgerows.

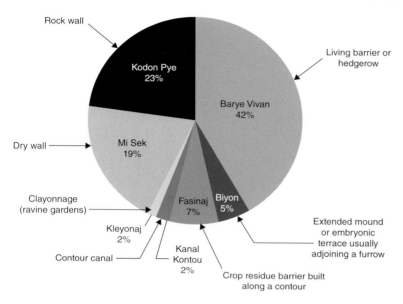

Figure 3.8 Soil conservation practices in Haiti. (*See insert for color representation of the figure.*)

Rice Production

Roughly half the world's population eat rice every day, and in many of the poorer countries including several of the SIDS, most of the population depends on rice as their
staple food. Table 3.7 shows the production of rice paddy in 25 of the SIDS for which
data are reported by the FAO (FAOSTAT, 2016). The major producers are Cuba,
Dominican Republic, Guinea-Bissau, Guyana, Suriname and Haiti – all producing
more than 100,000 tonnes of rice paddy a year.

In 20 of the SIDS, production has increased over the period 2003 to 2013, primarily
because of higher yields, which increased for the SIDS as a whole on average by 3.5%
annually over the period.

Cuba, the Dominican Republic and Guyana are the main producers. On six of the
islands, the area of land cultivated for rice decreased over the period. In Sao Tome and
Principe, cultivated area increased but yields went down, while on Mauritius the area
cultivated for rice increased substantially.

Because of its dependence on water, the production of rice is especially vulnerable to
the impact of climate change on patterns of rainfall and temperature. Irregular rainfall,
dry spells in the wet season, droughts and floods all have a negative impact on yields.
Sea-level rise increases the risk of flooding in coastal rice paddies and increases the
salinity of soils, which traditional rice varieties cannot tolerate.

A study of 227 irrigated rice farms in six Asian countries in 2004 estimated losses
due to climate change at 10–20% (Peng *et al.*, 2004). The 5th Assessment Report of the
IPCC is less certain about the reduction in yields for rice (Porter *et al.*, 2014); however,
a detailed 2009 report by IFPRI estimated yield reductions of 10–15% depending on

Table 3.7 Rice paddy production in 25 SIDS.

Small Island Developing States	2013 production, tonnes	2003–2013 annual growth in area	2003–2013 annual growth in production
Belize	20,505	5.5%	8.4%
Comoros	29,000	3.0%	4.6%
Cuba	672,600	2.8%	1.3%
Dominica	No data	−1.7%	−0.6%
Dominican Republic	824,000	2.0%	3.8%
Fiji	5000	−10.7%	−10.2%
Guam	No data	5.0%	7.4%
Guinea-Bissau	209,717	3.5%	4.7%
Guyana	823,800	3.2%	4.4%
Haiti	169,300	4.2%	4.7%
Jamaica	31	5.9%	6.4%
Maldives	No data	0.2%	2.8%
Mauritius	646	21.4%	12.3%
Micronesia Fed. states	165	0.1%	5.7%
Montserrat	No data	−1.7%	2.7%
New Caledonia	No data	−5.7%	−7.7%
Papua New Guinea	1300	3.2%	5.5%
Puerto Rico	No data	2.2%	1.5%
Sao Tome & Principe	No data	0.9%	−12.6%
Solomon Islands	4200	3.8%	4.5%
St Vincent & Grenadines	No data	−4.9%	2.2%
Suriname	262,029	1.9%	4.0%
Timor-Leste	87,000	0.2%	5.5%
Trinidad & Tobago	2859	2.4%	2.7%
Vanuatu	No data	−0.8%	−0.2%

Source: Data from FAOSTAT (2016).

the region. In the IFPRI study, developing countries also fare worse than developed countries (Nelson *et al.*, 2009). Rice yields in the Mekong Delta are projected to increase over the period 2010 to 2050 in the upper part of the basin in Laos and Thailand, and to decrease in the lower part of the basin in Cambodia and Vietnam (Mainuddin *et al.*, 2012).

The conflicting assessments and forecasts highlight the marked regional differences in the projections of rainfall and temperature over the next few decades. However, the immediate impact of climate change on the cultivation of rice is more related to the frequency of extreme weather, rather than to the longer-term trends in temperature, rainfall and CO_2 concentrations. Flooding of rice fields in coastal areas will become more frequent, and the salinization of soil much more of a problem. The FAO has promoted the use of salinity-tolerant rice varieties, and has collaborated with agricultural research institutes in Vietnam to develop high-yielding rice varieties with good tolerance of salinity for planting in saline-affected soils in the Mekong River Delta (FAO, 2008). Also in Vietnam, structural changes for better water control have become increasingly necessary because of the higher risk of flooding, and farmers have had to shift practices in order to live with floods rather than trying to control them (Chinvanno *et al.*, 2008).

Cropping patterns, planting timings and farm management techniques have all been adjusted to adapt to the threat of climate change. Embankments have been built to protect rice farms from floods, and new drought- and submergent-tolerant varieties of rice are being produced and made available (FAO, 2010a; Kawasaki, 2010). Many farmers are diversifying their production, growing other cereals, vegetables, and combining rice with aquaculture. If upland areas are available, farmers will spread the risk of climate change impacts by bringing them into production. For example, in the Na Dok Mai village in Thailand, villagers estimated that about 10% of the families lose their entire rice crop each year due to floods. However, some 80% of the villagers have rice paddies in both the floodplain and on higher ground. So even if the floodplain rice is flooded, the crop on higher ground can be safely harvested (Kawasaki, 2010).

The New Rice for Africa initiative (NERICA) promotes upland rice cultivation, and the varieties developed under this programme have better tolerance to drought stress. High-yielding varieties that tolerate salinity have also been developed in Vietnam (FAO, 2008).

The cultivation of rice is also a major source of emissions of methane – accounting for 30% of total methane emissions in China in 1994, and 35% of all CH_4 emissions in India in 2006 (Leip and Bocchi, 2007). Fully half of the methane emitted in Thailand is due to rice cultivation (Kawasaki, 2010). Actions to improve water management in rice cultivation are critical for both adaptation to climate change and the mitigation of greenhouse gas (GHG) emissions from rice production practices. In East Asia, changes to water management for paddy rice over the last several years have substantially reduced CH_4 emissions. For example, a large part of the of the paddy (flooded) rice production area has shifted from being continuously flooded to being drained at mid-season, resulting in an average 40% reduction in methane emissions and an overall improvement in yield due to better root growth and fewer unproductive panicules (DeAngelo *et al.*, 2005).

On Bohol Island in the Philippines, the implementation of the alternate wetting and drying (AWD) approach resulted in more efficient water management and increased yields by 30% in a double-cropping system. The AWD practice therefore generates multiple benefits: reduced methane emissions, reduced water use, increased productivity, and improved food security (Bouman *et al.*, 2007).

Direct seeding in wet or dry soils as opposed to transplanting into flooded soils, and the recent introduction of permanent raised beds with furrow irrigation in India show promise. Fully aerobic upland rice is another adaptation response to reduced water availability. The development of high-yielding upland rice systems in Brazil's Cerrado using improved upland varieties and supplemental irrigation shows that the transition to high-yielding aerobic rice is possible (Pinheiro *et al.*, 2006).

While improved water management is beneficial in terms of reducing methane emissions, achieving net reduction in GHG emissions from rice requires an analysis of the complete production cycle, because the wetting and drying practices that reduce methane emissions increase nitrous oxide (N_2O) emissions – a potent greenhouse gas. Increased N_2O emissions depend on the extent of the area converted from flooded to drained cultivation, and the rate at which improved fertilizer management practices can be adopted to offset the potential increase in N_2O emissions (Padgham, 2009).

Irrigation

About three-quarters of the water used in developing countries is for agriculture. In many countries, water resources are already overtaxed, and the situation will worsen under climate change due to increased temperatures, changes in runoff volumes, and the severity of extreme weather. Increasing urbanization and industrialization, always on the rise in developing countries, will increase the demand for water and compete with water for agriculture – which generally has a lower economic value.

Climate change presents challenges that will require major improvements in the way water is collected and used for food production, especially where agriculture is dependent on seasonal rainfall. In almost all tropical areas, adaptation measures to capture and conserve more rainfall, switch to drought-tolerant crops, and establish supplementary irrigation will reduce some of the expected negative effects of climate change.

A key adaptation response to climate change impacts on agriculture is to make the transition from rainfed agriculture to irrigated agriculture. Irrigation severs the traditional link between agriculture and rainfall. It breaks the chains that bind rainfed agriculture to the seasonal rains, and it offers the possibility of much higher yields, because production can continue all year long.

Adaptation Technologies for Agriculture

There are many new and improved technologies that can help farmers cope with a changing climate. Table 3.8 summarizes the available options – many of which can and should be used in combination (Clements *et al.*, 2011).

Adapting agriculture to climate change requires a programme of activities that are linked and coordinated with the selected technologies. Activities planned for design and implementation should include (Ngigi, 2009):

- Technological innovations: improved crop varieties, early warning systems, improved land and water management, integrated pest management.
- Government subsidies: agricultural subsidies for agricultural support services to cushion farmers against the impacts of climate change.
- Farm production practices: farm production, land use, land topography, irrigation, and timing of operations.
- Farm financial management: crop insurance (in case of crop failure related to variations in weather conditions), crop shares and futures, income stabilization programmes, and diversification of incomes.

Table 3.8 Adaptation technologies for agriculture.

Technology category	Adaptation technologies
Planning for climate change and variability	• National climate change monitoring system • Seasonal to interannual forecasting • Decentralized community-run early warning systems • Climate insurance
Sustainable water use and management	• Sprinkler and drip irrigation • Fog harvesting • Rainwater harvesting
Soil management	• Slow-forming terraces • Conservation tillage • Integrated soil nutrient management
Sustainable crop management	• Crop diversification and new varieties • New varieties through biotechnology • Ecological pest management • Seed and grain storage
Sustainable livestock management	• Selective breeding via controlled mating • Livestock disease management
Sustainable farming systems	• Mixed farming • Agro-forestry
Capacity building and stakeholder organization	• Farmer field schools • Community extension agents • Farmer to farmer exchanges • Forest user groups • Water user associations

Source: Adapted from Clements *et al.* (2011). Reproduced with permission of UNEP.

Focusing on West Africa, Ngigi (2009) documented several successful adaptation strategies for smallholder farmers:

• The use of shallow wells and hand-dug wells to supplement the shortfall in water for dry-season irrigation.
• The use of soil moisture improvement techniques such as *zai*, semi-moons and mulching.
• A more efficient use of water by using drop irrigation and the selection of high-yielding and high-value crops.
• The use of drought-resistant crop varieties and the improvement of on-farm irrigation efficiency through the use of better water application technologies.
• Bunds, agroforestry, crop rotation and rainwater harvesting have all been effective adaptation strategies.
• Agricultural diversification such the integration of livestock and crops (mixed farming) has also been practised in some areas with good results.
• The alternative use of waste water for irrigation in peri-urban irrigation schemes.

- Migration from drier to wetter regions to find wetter and more fertile lands (not always possible).
- Greater engagement in off-farm activities that generate revenue.

Persuading farmers to adopt new farming practices is not always easy. Farmers are generally poor, and poor people need to be very careful when trying a new approach because, if it doesn't work, they have very little in the way of resources to fall back on. It is entirely rational that farmers are risk-averse.

New techniques, innovative practices, and new varieties of crops that are drought-resistant and more resilient in the face of climate change impacts will gain acceptance only after well-organized, painstaking and continuous efforts to convince farmers of their worth. In Haiti, the mechanisms that have proved effective include:

- *Farmer field schools (FFS)*: where a selected farmer or group of farmers adopt the new technique and show the results. The farmers must be capable of explaining how they achieved the results and the comparative advantages of the technique. Farmers from communities in the area are brought to the FFS to spend a day seeing the results for themselves and learning how to apply the practice.
- *Farmer to Farmer (F2F) exchanges*: where groups of farmers exchange experiences and compare results. Group A visits Group B, which has adopted a new approach and can demonstrate good results. Then Group B goes over to Group A and works with them to show them how to get started. Group B may return when the crop is harvested, if there are new techniques that reduce post-harvest losses.
- *Seed producer groups*: where a group of farmers produces improved varieties of seeds within the Commune so that the seeds are available at a reasonable price to all the farmers in the area. This production site can be based at a farmer field school.
- *Seedling nurseries*: which need to be established within the Commune and managed by members of the community. The sale of the seedlings should not be subsidized. The seedlings should obviously be from varieties that are drought-resistant or resilient in other ways to the impact of climate change.
- *Grafting programmes*: where a group of master grafters is established in the Commune. Once trained and provided with a set of tools, master grafters both organize training days at their farms and hire out their skills to other farmers. Once again, it would be more effective if the master grafter is located close to a farmer field school so that several skills can be transferred to new farmers at the same time.

Unless these activities are included in the action plans and budgeted for over a sufficient period of time, the adoption of innovative practices will be extremely slow and in many instances fail completely (Bush, 2014).

Water Resources Management

One of the most commonly used measures of water availability is based on per capita metrics. Countries are classified as:

- water scarce if fewer than 1000 cubic metres of renewable freshwater is available per person per year;
- water stressed, if there are between 1000 and 1700 cubic metres available per person per year.

Approximately 2 billion people are currently living in areas faced with water stress or scarcity (PAI, 2011). A situation of water scarcity affects all social and economic sectors and threatens the health of many ecosystems that provide services to local communities.

The most water-stressed areas are typically those with few water resources, high population densities and high population growth rates. For instance, most of the countries in the Middle East and North Africa, the MENA region, cannot meet their current water demand. Seven of the world's ten most water-scarce countries are in the MENA region, and four of the top ten are also small island developing states: Bahamas, Bahrain, Maldives and Singapore. The predictions are that freshwater stress for islands will get worse (Karnauskas *et al.*, 2016).

Water is the key to survival. If water resources are not managed effectively, small island developing states will be the first to suffer. Water resources cannot be managed piecemeal, particularly on small islands where the linkages between watersheds and the ocean environment are so closely related.

Many small island developing states are already experiencing varying degrees of water-related problems.

- Water supply in Comoros on the islands of Grande Comoros, Moheli and Anjouan is threatened by salt intrusion into many of the coastal boreholes.
- In the Seychelles in 1998, water shortages were so severe that brewing and fish-canning industries were forced to close.
- In Bermuda and the US Virgin Islands, all new buildings are required to harvest enough rainwater to serve their residents.

These measures point to the need to manage water resources in a coordinated, integrated and effective manner. Where SIDS differ from other larger countries is in the immediacy of their problems and their limited capacity to respond effectively. With relatively small land mass and limited water resources, the pressures of economic development coupled with climate change make water shortages, flooding, soil erosion and salination a present and worsening threat for all water-use sectors. These pressure and demands are now close to exceeding the carrying capacity of many of the islands, especially those with major urban centres and high population densities.

Integrated Water Resources Management (IWRM)

Integrated water resources management is an approach that aims to coordinate the sustainable development and management of water resources in order to maximize the efficiency of water use. The IWRM approach for small island developing states has typically been based on the following principles (UNEP, 2012):

- It should be spatially conceptualized within a watershed and its receiving waters, from ridge to reef; also called Hilltops 2 Oceans (H20).
- It requires an understanding of the relationship between activities on land and coastal waters.
- It is based on an ecosystem approach.
- It is adaptive, interactive, and entails a process of balancing different goals and priorities at national, watershed and community levels.
- It is issue-based with defined 'entry-points'. These can be at national, watershed or community levels.

Table 3.9 Lack of effective water resources management in SIDS.

1	There is a lack of integration between sectoral water-related policies. This leads to fragmented programmes and inefficient utilization of technical capacities and financial resources.
2	There is a lack of decentralization and effective local government agencies, coupled with low capacity at the local level, which undermines effective IWRM action at the grassroots level.
3	Many SIDS have traditional or customary systems of land tenure, poor land use practice, and inadequate land-use policies, all of which can have a negative impact on the management of water resources. Traditional values, beliefs and rights can pose significant impediments to the effective management of water resources.
4	There is often a deep-seated belief that water is a free good, and that the provision of water is a social service. Water is not seen as a valuable resource to be conserved and carefully managed.
5	There is little or no coordination between the different authorities concerning data collection, analysis and knowledge management. The result is inefficient use of technical and financial resources. Decisions are taken without full knowledge of the situation because data are lacking or have not been analyzed and shared.
6	Sound water management strategies are not affordable by many SIDS. Governments have budget constraints, and because water management is spread among several agencies and ministries, insufficient funds are allocated for the management of water resources.
7	Legislation that should regulate the management and use of water is inadequate or non-existent. The 'polluter pays' principle cannot be applied and enforced without laws that authorize its application. The law should clearly define which ministry is responsible for which part of water resources management and how ministries should coordinate and collaborate.

Source: Adapted from UNEP (2012). Reproduced with permission.

- It focuses on incremental steps and tangible issues.
- It focuses not on outcomes, but on the practical activities involved in achieving the outcomes.

But many small island developing states have been slow to formulate and implement water resources management approaches. The reasons for this situation are summarized in Table 3.9.

The Caribbean SIDS

The Caribbean islands are struggling to find enough water for their residents and for essential economic sectors like tourism, industry and agriculture. In many of the islands the annual per capita freshwater availability is far below the $1000\,m^3/yr$ benchmark commonly used to define scarcity.

In most Caribbean SIDS, water comes mainly surface flows – rivers, springs and ponds – and from groundwater resources. In some of the smaller islands, rainwater is harvested. Desalination technologies are increasingly being installed in islands where the demand for fresh water is impossible to meet from natural rainfed sources or groundwater. Desalination is an expensive option, which gives some indication of how bad the scarcity of water is on many of the islands (GWP, 2014).

The Pacific SIDS

The Pacific SIDS vary immensely in size, hydrology and geology. Large and high volcanic islands like Papua New Guinea are at one end of the scale; at the other are the much smaller coral atolls like Kiribati and the Solomon Islands which are scarcely above sea-level. The tiny island state of Niue has no natural surface water and is entirely dependent upon rainwater harvesting and groundwater. IWRM strategies obviously differ according to the local supply and demand situation.

Much like the Caribbean, integrated water resources management has not been widely applied among the Pacific SIDS. On many islands, there is clearly a lack of technical capacity and know-how. A recent report (UNEP, 2012) identified the problems as follows:

- Insufficient knowledge of water resources distribution, flow and management (hydrology, hydrogeology and recharge rates).
- Insufficient education, training and capacity in IWRM practice and water use efficiency at various levels, including government, private sector and community.
- Lack of access to and awareness of appropriate technologies and methodologies for IWRM and water use efficiency, including wastewater management and sanitation.
- Lack of access to models and demonstrations of IWRM and Water Use Efficiency at national and catchment level that are appropriate to small island states.
- Inappropriate policy legislation, planning and administration.

The Indian and Atlantic Ocean SIDS

Referred to as the AIMS group of SIDS, the group includes Cabo Verde, Comoros, Maldives, Sao Tome et Principe, and the Seychelles. Several of the islands have developed as upmarket tourism destinations that require large quantities of water. Agricultural requirements are often subordinated to the demands of the tourism industry. Cabo Verde and Mauritius have invested in water supply and wastewater treatment infrastructure. Treated wastewater can be recycled for agriculture – a technology that has great potential.

Most of the SIDS in the Indian and Atlantic Oceans have some sort of water management structures in place, and the more developed islands like Mauritius and Cabo Verde have water resources monitoring systems in operation and possess a degree of technical capacity.

Overall, the progress made in integrated water resources management in the SIDS over the last decade has been limited. This must change if climate change is to be successfully managed.

Agricultural Water Management (AWM)

Because of the changing climate, managing water for agriculture is a key adaptation strategy for smallholder farmers. AWM includes crop husbandry, soil and water conservation, rainwater harvesting and management, irrigation and drainage, wetlands management, and all aspects of land and water management. Water from different sources – rainwater, surface water, groundwater and wastewater – can be used for crop and livestock production including aquaculture. AWM can be classified into several categories (Ngigi, 2009):

- soil and water conservation
- runoff harvesting and management
- on-farm storage for supplementary irrigation
- runoff diversion and spreading
- spate irrigation (flood diversion and spreading on bunded plots)
- wetlands farming – valley bottoms or flood recession cultivation
- stream diversion for smallholder irrigation using either gravity or pumps
- irrigation technologies (low-head drip, sprinkler, furrow and basin, micro systems)
- soil management and fertility improvement
- conservation agriculture (conservation tillage, crop residue management, agroforestry, etc.).

Rainwater Harvesting and Management

In arid and semi-arid regions where rainfall is generally inadequate, it makes sense to capture as much as possible and to use this water efficiently. Moisture conservation employing conservation agriculture and/or the construction of rainwater management (e.g. storage in farm ponds, water pans, sand/subsurface dams, earth dams or tanks) are practices that are becoming increasing widespread.

The collection and storage infrastructure can be natural or constructed. Examples include:

- Below-ground tanks or cisterns, either lined or unlined, into which rainwater is directed from the ground surface. Volumes are typically small and they are usually used by one household or institution – for instance, a school or clinic.
- Small reservoirs with earth bunds or embankments to contain runoff or river flow. They are typically built from soil excavated from the reservoir to increase storage capacity. A spillway or weir allows controlled overflow when the storage capacity is exceeded. Reservoir areas range from 5 to 12 hectares. In Ghana, the mean storage volume was estimated at about 50,000 m^3 (Elliot *et al.*, 2011).
- Groundwater aquifers can be recharged by directing water down an unlined well. Groundwater recharge is also an added benefit of unlined reservoirs: stored water will infiltrate permeable soils during storage and eventually reach the water table.
- Many runoff control methods for irrigation incorporate inundation or extended contact time with soils to increase topsoil moisture. Many traditional methods have been used for centuries. Examples include variations of contour farming, which slows runoff, decreases erosion and increases infiltration.

Farmers in Africa are adopting a variety of rainwater harvesting and management practices. In Kenya, farm ponds have reportedly significantly increased crop production and farmers' incomes. In Ethiopia, lined farm ponds have been widely adopted after the government subsidized the cost of the UV-resistant plastic liner – the most expensive part of the pond (Ngigi, 2009). In Ethiopia, the ponds hold between 40 and

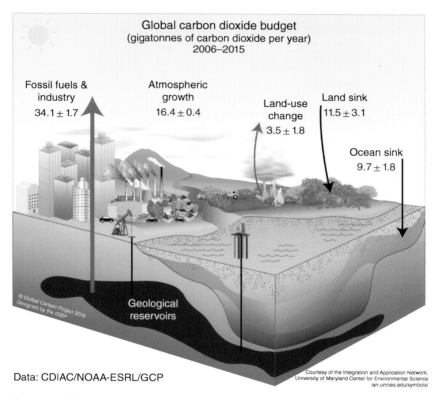

Figure 1.3 Global carbon dioxide budget over the period 2006 to 2015. *Source:* LeQueré *et al.* (2016), www.earth-syst-sci-data.net/8/605/2016/. Used under CC-BY-3.0.

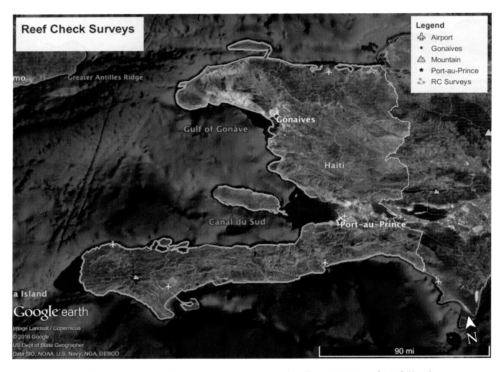

Figure 2.3 Reef Check surveys of Haiti. *Source:* Reproduced with permission of Reef Check International.

Figure 2.4 A typical former coral reef at La Gonave in Haiti overfished to the point where it has become an algal dominated reef with a few stubs of coral surviving but no fish. *Source:* Reproduced with permission of Reef Check International.

Figure 2.6 The island of Malé in the Maldives. *Source:* https://en.wikipedia.org/wiki/Mal%C3%A9#/media/File:Male-total.jpg. Used under CC BY-SA 3.0.

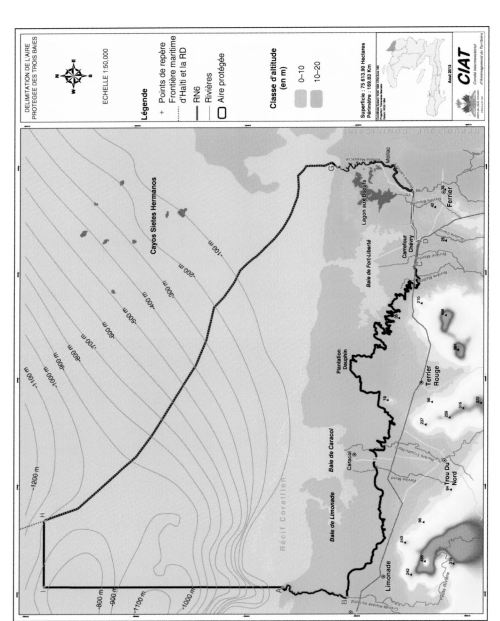

Figure 3.2 The limits of the Three Bays Protected Area.

Figure 3.3 Charcoal production from mangroves close to the PN3B. *Source:* Courtesy of Andy Drumm, Drumm Consulting.

Figure 3.4 Plastic bottles and floating garbage washed up on one of the small islands on the west coast of Haiti. *Source:* M. Bush.

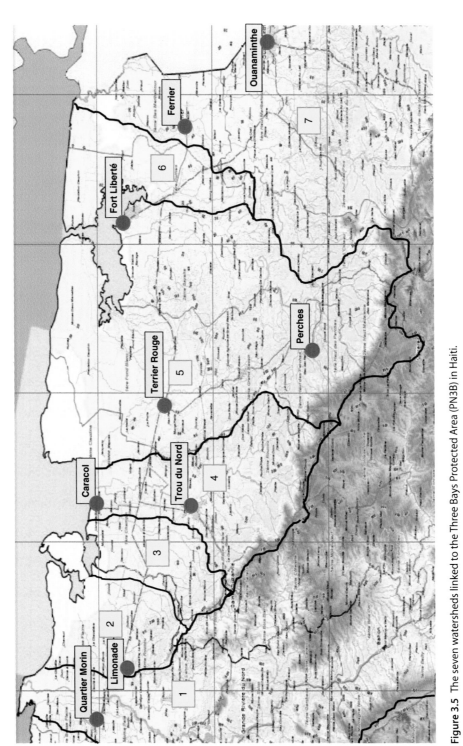

Figure 3.5 The seven watersheds linked to the Three Bays Protected Area (PN3B) in Haiti.

Spatial configuration of watershed impact-generating areas (IGAs), impact receiving areas (IRAs), and impact flows

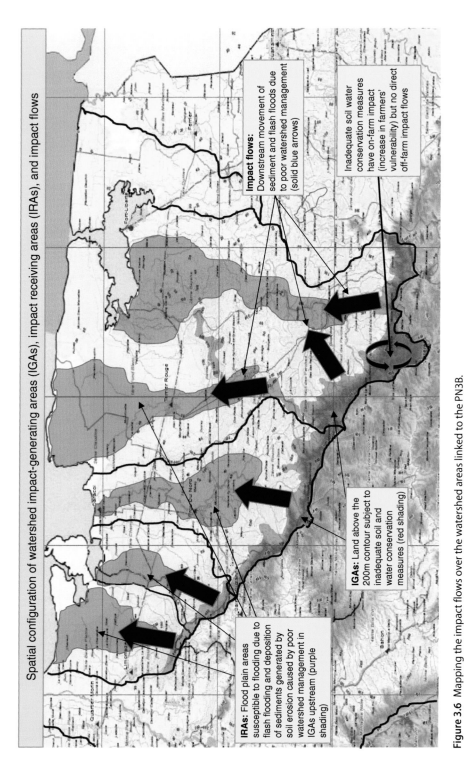

Impact flows:
Downstream movement of sediment and flash floods due to poor watershed management (solid blue arrows)

Inadequate soil water conservation measures have on-farm impact (increase in farmers' vulnerability) but no direct off-farm impact flows

IGAs: Land above the 200m contour subject to inadequate soil and water conservation measures (red shading)

IRAs: Flood plain areas susceptible to flooding due to flash flooding and deposition of sediments generated by soil erosion caused by poor watershed management in IGAs upstream (purple shading)

Figure 3.6 Mapping the impact flows over the watershed areas linked to the PN3B.

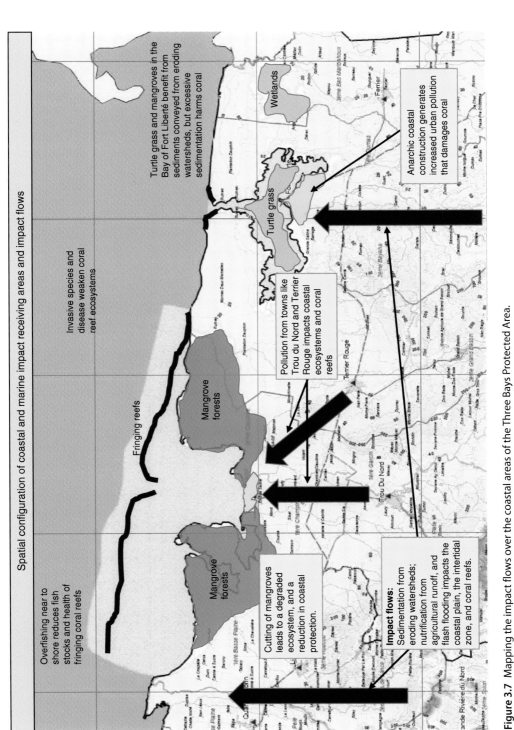

Figure 3.7 Mapping the impact flows over the coastal areas of the Three Bays Protected Area.

Spatial configuration of coastal and marine impact receiving areas and impact flows

Turtle grass and mangroves in the Bay of Fort Liberté benefit from sediments conveyed from eroding watersheds, but excessive sedimentation harms coral

Wetlands

Anarchic coastal construction generates increased urban pollution that damages coral

Invasive species and disease weaken coral reef ecosystems

Turtle grass

Pollution from towns like Trou du Nord and Terrier Rouge impacts coastal ecosystems and coral reefs

Fringing reefs

Mangrove forests

Mangrove forests

Overfishing near to shore reduces fish stocks and health of fringing coral reefs

Cutting of mangroves leads to a degraded ecosystem, and a reduction in coastal protection.

Impact flows: Sedimentation from eroding watersheds; nutrification from agricultural runoff, and flash flooding impacts the intertidal zone, and coral reefs.

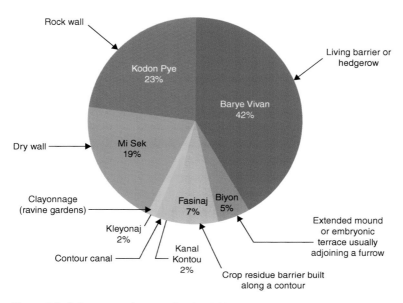

Figure 3.8 Soil conservation practices in Haiti.

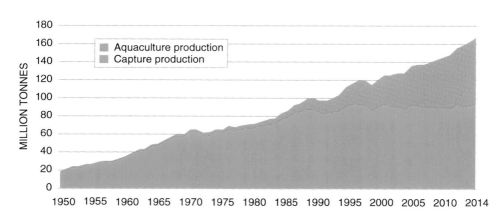

Figure 3.9 Aquaculture production since 1950. *Source:* FAO (2016). Reproduced with permission.

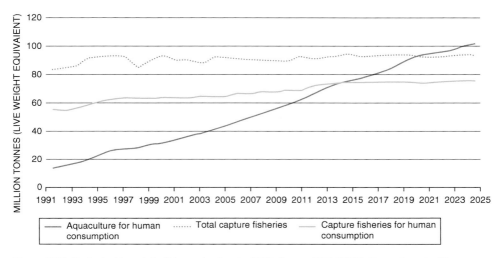

Figure 3.10 Projected trends in fish production to 2025. *Source:* FAO (2016). Reproduced with permission.

Figure 4.6 The inauguration of the Les Anglais photovoltaic hybrid minigrid system on 1 June 2015. Photo: EarthSpark, used with permission.

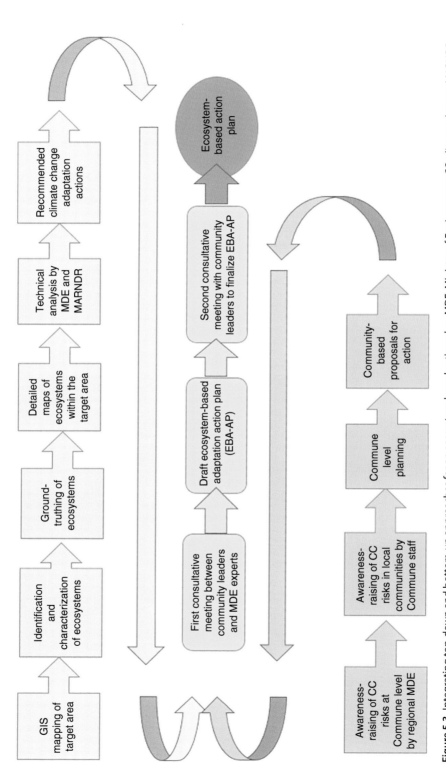

Figure 5.3 Integrating top-down and bottom-up approaches for an ecosystem-based action plan. MDE, Ministry of Environment; CC, climate change; MARNDR, Ministry of Agriculture.

$60\,m^3$ of water. If the pond is management by a group of farmers, the volume may be greater: up to $125\,m^3$ of water is being stored in ponds in Rwanda.

Evaluation of farm ponds in Kenya showed a payback time of 4 years for a $50\,m^3$ pond costing US$650. Revenue increased by US$150 per season with two crops a year being produced. Although certainly useful, rainwater management systems cannot ensure sustainable production. The need for improved agronomic practices, such as timely planting to take advantage of expected rainfall, are essential. The challenge is to convince risk-averse smallholders that they can benefit by improving their agricultural production systems.

Another effective technology employed in several African countries is *zai*. These are pits about 60 cm in diameter and 30 cm deep, spaced out at roughly 1 m intervals, giving between 10,000 and 15,000 pits per hectare. In each pit, between 5 and 9 plants, mainly sorghum, millet and maize, are grown. Although requiring initially more labour, the practice is simple, easy for poor farmers understand, and effective. Manure can also be incorporated into the pits to improve soil condition and fertility. *Zai* is both a soil moisture conservation measure and a soil fertility improvement technique. They are particularly effective on degraded soils.

Farmers also use stone contour bunds to reduce the speed of runoff, allowing infiltration into the *zai* which collect and concentrate the runoff. The larger the planting pits and the wider the spacing, the more water can be harvested from the uncultivated micro-catchment areas between the pits. Originating in West Africa, the *zai* system has now spread to East and southern Africa – testament to its effectiveness.

Rainwater Harvesting from Rooftops

The collection of rainwater from rooftops is an ancient practice, but one still useful for supplementing household water supply. The increased proportion of metal or tile roofs and the easy availability of plastic piping have decreased the cost of household rainwater harvesting (RWH).

RHW is widely practised in many countries. Over 60 million people were using RWH as their main source of drinking water in 2006, and that number is projected to increase to more than 75 million by 2020. In most developing countries, RWH is used to collect water for potable and other household uses. In regions with reliable potable household water supply, rainwater is often collected for landscape irrigation (lawns and gardens), toilet flushing, and other uses. The basic elements consist of the roof to intercept the precipitation, the gutters and pipes that carry the runoff, and the covered ground-level storage reservoir. Water quality can be protected by adding one or more of the following: filtration and screening, chemical disinfection, or a 'first-flush' system, which discards the initial volume of rooftop runoff.

A first-flush separation system is essential. It prevents dirt, leaves, bird droppings, and other potentially harmful and contaminating materials from entering the water collection container. Many collection systems therefore incorporate a simple first-flush system that diverts the initial flow from the collection surface into a separate tank. As a rule of thumb, contamination is halved for each mm of rainfall discarded (Elliot *et al.*, 2011). The contaminated water can still be used for irrigating family or community plots where the quality of the rainwater is not so important (PA, 2016).

Rainwater harvesting is a useful adaptation approach even in developed countries. Even if safe, reliable piped supplies are available, RWH for non-potable uses can partially offset increases in household use. In some parts of the US, half of all residential and institutional water use goes to landscape irrigation; simple rain barrels are commonly used to water landscapes without using the piped water supply. A third of residential water in Europe is used for flushing toilets and another 15% for running washing machines and dishwashers. In Germany and elsewhere, the use of rainwater for these non-potable uses is now increasingly common (Elliot *et al.*, 2011).

In the Caribbean, climate change is likely to make rainfall more unpredictable and less frequent. It therefore makes sense to install a storage system that buffers the intermittency of the supply and provides water during dry spells.

Rainwater harvesting in the Caribbean is common. The technology is widespread in the region: Antigua and Barbuda, the Bahamas, the Virgin Islands (both US and British), St Vincent and the Grenadines (including Grenada) have all used rainwater harvesting for decades. Although rainfall varies across the Caribbean, it is highly seasonal, with more than 80% of annual rainfall often occurring within a few months of the year. It therefore makes sense to capture and store rainwater during the rainy season and to use it during the months when rainfall is much less frequent and the amount inadequate. In Jamaica, it has been estimated that more than 100,000 people rely to some extent on rainwater – especially those in rural areas without piped supplies of potable water. Some hotels in Jamaica collect rainwater as a supplementary water supply – an approach that can save up to a third of water utility costs. On the island of Cariacou (one of the Grenadines) rainwater is the only supply of water, so rainwater is used for everything from drinking to irrigation. And in the Bahamas, rainwater is used as a supplementary supply for large apartment buildings, hotels and restaurants (GWP, 2016).

The quality of rainwater is excellent: it generally has little or no contamination even in urban areas. It is easily contaminated, however, during the process of being collected. Catchment surfaces should be kept clean. Rainwater tanks should be designed to protect the water from contamination by leaves, dust, insects, vermin, and other pollutants. Tanks should be sited away from trees, with good fitting lids and kept in good condition. Incoming water should be filtered or screened.

Algae will grow in the tank if sunlight can enter. Keeping the tank dark and cool will usually prevent algae growth. Mosquito-proof screens should be fitted to all openings.

For small island states with extensive tourism infrastructure, it makes sense to examine rainwater harvesting to reduce water use in hotels and restaurants. The supply of potable water in small islands should never be taken for granted, and simple rainwater harvesting technologies that reduce consumption should be mandatory on all new buildings, Potable water should never be used for outside non-potable uses: landscape watering, golf courses and lawns.

Wastewater Treatment and Recycling

The term 'wastewater' explains why water that can so easily be cleaned up and recycled is so often treated as worthless, when in fact it is a valuable resource. Essentially all the water from households, including sewage, can be treated and recycled. Its value stems from the fact that it can be used for agriculture, and that the treated water contains elements (nitrogen, phosphorus and potassium) that are essential for plant growth. The term 'wastewater' is a misnomer: treated urban wastewater is a valuable liquid fertilizer.

Few small island states have treatment plants that process all the wastewater from major towns and cities. The tendency has always been to pump the wastewater out to sea – the cheapest option. But not only is this a huge waste of a valuable resource; it is damaging to coral reefs and the marine environment. By maximizing the collection and treatment of urban wastewater, not only can peri-urban agriculture be better protected against drought, but coastal and marine ecosystems are also less stressed and more likely to withstand the impacts of climate change (higher seawater temperatures and increased acidity).

The world's population will increase substantially over the next decades, and with that comes an increasing demand for water to meet all the essential needs of modern societies. Linked to this will be a commensurate increase in the amount of wastewater generated from these uses of water. Many communities throughout the world are close to, or have already reached, the limit of their supplies. On small tropical islands, the situation is even more critical since the supply of fresh water on many islands is already inadequate to meet the demand – particularly where tourism is on the increase.

Growing urbanization in water-scarce areas of the world makes these problems a lot worse. The demand for water for domestic, industrial, commercial and agricultural use is set to increase substantially, and the management of the wastewater arising from these uses is a problem that intensifies at the same rate. Wastewater cannot continue to be dumped in the sea or the nearest river. Apart from the fact that wastewater is a valuable resource, the impact on coral reefs and marine ecosystems trying to adapt to increased water temperatures and higher acidity is a major concern.

The re-use of treated wastewater in agriculture is an option that has increasingly been adopted in regions of water scarcity. Many regions of the world are experiencing growing water stress resulting from the constant increase in the demand for water where the supply is static or decreasing and droughts are frequent. Climate change will increase these pressures: global warming of 2 °C, now practically certain, could lead to a situation where 1 to 2 billion more people may no longer have enough water to meet their basic needs: consumption, hygiene and food.

Water stress is also caused by pollution from wastewater and runoff from expanding cities, much of it only partially treated, from the release of agricultural fertilizer, and from the contamination of aquifers. This pollution causes eutrophication of surface waters frequently resulting in algal blooms, rendering surface water unfit for consumption. Pollution can reduce dissolved oxygen levels to the point where the water harms fisheries and marine ecosystems.

The scarcity of water often has substantial economic, social and political costs. The drought in Kenya in 1998–2000 is estimated to have reduced GDP by 16%. Water scarcity in diverse regions such as the western US, Australia and northern China is counted in the hundreds of millions of dollars.

When the demand for water cannot be met from existing sources, water management agencies tend to divert water from farmland to cities, since water has a higher economic value for urban and industrial uses than for most agricultural purposes. In this situation, the re-use of treated wastewater for agriculture enables fresh water to be used for more economically and socially valuable purposes, while providing farmers with a reliable and nutrient-rich source of water. This substitution also has potential environmental benefits by reducing the amount of wastewater effluent discharged downstream or out to sea. Well-managed and effective wastewater re-use projects therefore offer a triple benefit: for urban users, farmers, and the environment.

There are several thousand water reclamation facilities operating worldwide. The treated water is used for a variety of uses: irrigated agriculture, urban landscaping, recreation, industrial cooling and processing, and groundwater recharge. The majority of these facilities are in Japan, the US, Australia and Europe, but similar treatment plants can be found in many other countries.

It is estimated that within the next few decades more than 40% of the world's population will live in countries facing water stress or water scarcity. Growing competition between urban and agricultural uses of high-quality freshwater supplies, particularly in arid, semi-arid and densely populated regions, will ramp up the pressure on this increasingly scarce resource.

For many farmers, wastewater may be a more reliable source of water than other more conventional sources, but this depends on the supply to urban areas also being reliable. The value of recycled water has long been recognized by farmers not only as a water resource but also for the nutrients it contains, elements that promote plant growth and for soil conditioning. The total land irrigated with raw or partially diluted wastewater is estimated at 20 million hectares in 50 countries (FAO, 2010b). This is about 10% of total irrigated land. Recycling and re-use of wastewater can also relieve pressure on water resources caused by pumping water from groundwater and aquifers.

In Europe, most of the re-use systems are located on coastal areas and the islands of the semi-arid regions of the Mediterranean and in highly urbanized areas where water is becoming scarce. In some areas water resources have fallen below the chronic water scarcity level of $1000\,m^3$ per person per year. Long distances between water sources and users also cause serious regional and local water shortages, and water scarcity worsens with the influx of summer tourists to the Mediterranean coasts.

Some European countries have issued guidelines and regulations on wastewater reclamation and re-use. Water re-use is specified as an option in the European Water Framework Directive (WFD) which emphasizes the need to integrate health, environmental standards, the provision of services and financial regulation to achieve greater efficiency in the overall use of water resources.

Reclaimed Water for Agricultural Use

Wastewater has been used for agriculture for several thousand years: the Greeks and the Romans were among the first to recognize that water could be used more than once. Because agriculture uses 70–80% of water resources, when water is scare or supplies are interrupted, farmers are often turned to wastewater as an alternative source. While recycled water is a relatively small component of water use overall, in some countries it plays a prominent and essential role, especially for agriculture. In Kuwait, for example, re-used water accounts for 35% of total water extraction. Globally some 525,000 hectares are irrigated with reclaimed water.

Concerns about the health risks associated with the use of recycled water are often expressed. What is clear is that the practice must be carefully controlled and monitored, just as carefully as if the water was intended for urban households. The first regulations were drafted in California in the US; now the 2006 WHO guidelines are the accepted reference. The guidelines set out the level of wastewater treatment required, crop restrictions, wastewater application methods, and the control of human exposure. The health-based targets used by the WHO apply to a reference level of acceptable risk, defined in terms of Disability Adjusted Life Years or DALYs. Depending on the circumstances, various health protection measures are possible, including waste treatment, crop restriction, adaptation of irrigation technique and application time, and control of human exposure. The hazards associated with the consumption of wastewater-irrigated produce include excreta-related pathogens and some toxic chemicals. The risk from infectious pathogens is significantly reduced if food is cooked rather than eaten raw (EPA, 2004).

The following health protection measures, generally called barriers, each add another layer of protection for consumers:

- wastewater treatment
- crop restrictions
- wastewater application techniques that minimize contact (e.g. drip irrigation)
- withholding periods that allow time for pathogens to die off naturally
- hygienic practices at food markets and during food preparation
- health and hygiene promotion
- washing produce, disinfection and cooking.

The highest quality recycled wastewater is achieved by dual membrane and reverse osmosis technologies, but these are expensive systems and better suited for high-value cash crops or the recharge of aquifers. The level of treatment and the standards that apply depend on the agricultural produce – whether it will be consumed, and if so, whether raw or cooked.

Table 3.10 provides some guidelines on the level of treatment required for certain types of agricultural production.

Table 3.10 Examples of types of crops irrigated with treated wastewater.

Type of crops	Examples	Treatment requirements
Field crops	Barley, maize, oats	Secondary + disinfection
Fibre and seed crops	Cotton, flax	Secondary + disinfection
Vegetable crops that can be consumed raw	Tomatoes, lettuce, melon, cucumber, cabbage	Secondary + filtration + disinfection
Vegetable crops processed or cooked before consumption	Onions, potatoes, hot pepper, sugar beet, sugar cane	Secondary + disinfection
Fodder crops	Alfalfa, barley, cowpea	Secondary + disinfection
Orchards and vineyards	Apricot, orange, peach, plum, grapes	Secondary + disinfection
Nurseries	Flowers and horticulture	Secondary + disinfection
Commercial woodlands	Timber, fuelwood, charcoal	Secondary + disinfection

Table 3.11 Levels of wastewater treatment and systems.

Treatment level	Treatment
Preliminary	Raw effluent is screened to remove large solid materials and grit
Primary	Sedimentation tanks allow solid material to settle, while oil and grease rise to the top. Floating material is removed for separate treatment.
Secondary	Wastewater flows into an aerated stirred tank where microbiological processes remove dissolved organic material. The liquid is then clarified and the sludge is removed for separate treatment. The sludge is often disposed of as waste, but in fact has value as a fertilizer.
Tertiary	This level is intended to remove pollutants such as nitrogen and phosphorus or specific industrial pollutants. The wastewater may be chlorinated to kill the remaining bacteria.

Wastewater Treatment

Municipal sewage treatment typically involves the processes outlined in Table 3.11 (FAO, 2010b).

For wastewater use in agriculture, secondary treatment is mandatory and disinfection is recommended when agricultural produce will be eaten raw. The removal of nitrogen and phosphorus is not carried out since these elements are essential nutrients for plant growth; the treated wastewater therefore has significant value as liquid fertilizer.

Discharging inadequately treated wastewater to the sea or to surface waters can cause significant pollution and eutrophication. The re-use of wastewater for agriculture therefore brings tangible and measurable environmental benefits. Reclaimed water can also be used to restore wetlands, augment groundwater flows and recharge aquifers. The type of irrigation system to be employed depends on several factors. Table 3.12, based on FAO guidelines, sets out the factors influencing the choice of irrigation systems and the levels of protection recommended (FAO, 2010b).

Local geography is an important factor for the feasibility of recycling water. The source of reclaimed water needs to be in reasonable proximity to the intended user to minimize the need for sewerage, piping and pumps.

Fisheries and Aquaculture

Fisheries and aquaculture play an essential part in maintaining global food security, especially in small island developing states. Approximately 130 million tonnes of fish are produced annually from marine and freshwater capture fisheries and from aquaculture, and are used directly for human consumption. Fish provide roughly 4.3 billion people with about 15% of their average intake of animal protein. In addition, about 10% of the world's population, predominantly from developing and emerging countries, rely heavily on fisheries and aquaculture for the income needed to buy food (IUCN, 2016).

In recent decades, total annual production from capture fisheries has levelled off at about 90 million tonnes because most marine resources are now fully exploited, and in

Table 3.12 Factors affecting the choice of irrigation system and special measures required for reclaimed water re-use.

Irrigation method	Factors affecting choice	Special measures for irrigation with reclaimed water
Flood irrigation	• Lowest cost • Exact levelling not required • Low water use efficiency • Low level of health protection	Thorough protection of field workers, crop handlers and consumers required by using protective equipment and clothing
Furrow irrigation	• Low cost • Levelling may be needed • Low water use efficiency • Medium level of health protection	Protection of field workers required, possibly of crop handlers and consumers by using protective equipment
Sprinkler systems	• Medium to high cost • Medium water use efficiency • Levelling not required • Low level of health protection, because of possible aerosol contact and dispersion	• Minimum distance 50–100 m from houses and roads • Water quality restrictions apply (i.e. pathogen removal) • Anaerobic wastes should not be used due to odour nuisance
Subsurface and drip irrigation	• High cost • High water use efficiency • Higher agricultural yields • Highest level of health protection	No additional protection is required but filtration may be needed to prevent clogging of drip feed pipes

many cases overexploited. But the rapid development of aquaculture has ensured that the demand for fish has been met. In fact, the rate at which aquaculture production has increased has enabled global fish supply to outpace population growth for the last 50 years, increasing at an average rate of just over 3%. Figure 3.9 shows the trends for capture and aquaculture production since 1950.

With capture fishery production relatively flat since the late 1980s, aquaculture has been responsible for the impressive growth in the supply of fish for human consumption. Whereas aquaculture provided only 7% of fish for human consumption in 1974, this share increased to 26% in 1994 and 39% in 2004. China played a major role in this growth as it represents more than 60% of world aquaculture production (FAO, 2016).

In 2014, fish harvested from aquaculture amounted to 73.8 million tonnes: 49.8 Mt of finfish, 16.1 Mt of mollusks, and 6.9 Mt of crustaceans (FAO, 2016). In 2014, 25 countries recorded annual aquaculture production in excess of 200,000 tonnes.

Although capture fisheries will continue to be an important source of fish, particularly for coastal communities in small island developing states, most of the future demand for fish will come from aquaculture. The development and promotion of aquaculture is therefore a critically important adaptation response to climate change. Figure 3.10 shows the projected increase in aquaculture production as estimated by the FAO.

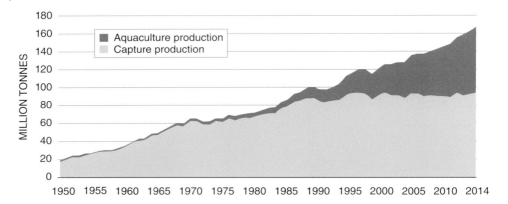

Figure 3.9 Aquaculture production since 1950. *Source:* FAO (2016). Reproduced with permission. (*See insert for color representation of the figure.*)

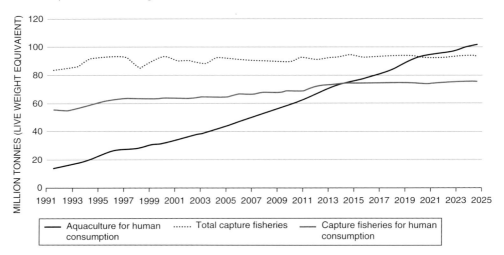

Figure 3.10 Projected trends in fish production to 2025. *Source:* FAO (2016). Reproduced with permission. (*See insert for color representation of the figure.*)

Continuing improvements in aquaculture related to feed formulation, feeding technologies, farm management and selective breeding have greatly increased the productivity of aquaculture. In addition, the reduced dependence on fish meal for feeds has decoupled marine fishers from aquaculture production, allowing more capture fish to be used directly for human consumption.

Climate change will have a range of impacts on aquaculture, but there are several adaptation measures that can effectively address climate variability.

- Aquaculture zoning to minimize risks (for new aquaculture) and relocation to less exposed areas for existing farms.
- Appropriate fish health management.
- Increasing efficiency of water use, water recycling, aquaponics, and so on.
- Increased feeding efficiency to reduce pressure and reliance on feed resources.

- Developing better-adapted seed stock which has tolerance to acidification, broader resistance to salinity, faster-growing strains and species.
- Ensuring high quality, reliable hatchery production to facilitate outgrow in more stressful conditions, and to facilitate rehabilitation of production after disasters.
- Improvement in monitoring and early warning systems.
- Strengthening farming systems including better holding structures (e.g. sturdier cages, deeper ponds) and management practices.
- Improving harvesting methods and value addition.

Some countries are already taking action to strengthen the resilience of aquaculture to climate change threats. In Vietnam, there are efforts to select for salinity-resistant catfish strains, and in Bangladesh the government is experimenting with salinity-resistant species, deepening aquaculture ponds, using depth-adjustable cages, and integrating fish farming with agriculture (FAO, 2016).

Disaster Risk Management

The consensus among climate change experts is that extreme weather events are likely to intensify and perhaps become more frequent. Higher sea surface temperatures result in greater evaporation and more water in the atmosphere. Higher ocean water temperatures fuel more intense and violent storms. The mechanisms are well understood, even if the location and timing of the storms are still unpredictable.

Few disasters, if any, can be attributed directly to climate change. But there is increasing evidence that many naturally-occurring catastrophic phenomena – hurricanes, cyclones, wild fires, heatwaves, and droughts among others – are intensified by global warming driven by climate change.

The Third UN World Conference on Disaster Risk Reduction was held in Sendai, Japan, in March 2015, just as Cyclone Pam was devastating the island of Vanuatu. Speaking at the conference, the President of Vanuatu, Baldwin Lonsdale, told the delegates that the damage on the Pacific island was unprecedented:

> *This is a major calamity for our country. Every year we lose 6% of our GDP to disasters. This cyclone is a huge setback for the country's development. It will have severe impacts for all sectors of economic activity including tourism, agriculture, and manufacturing. The country is already threatened by coastal erosion and rising sea-levels in addition to five active volcanoes and earthquakes.*
>
> (FAO, 2016)

According to a 2014 assessment of the period 2003 to 2013 conducted by the FAO, the agricultural sector – including fisheries and aquaculture – accounted for almost a quarter of the economic impact caused by medium- and large-scale disasters in developing countries. This sector is particularly vulnerable to disasters. Typhoon Haiyan in 2013 affected 40,000 fishers and damaged or destroyed 30,000 fishing vessels.

The 2004 tsunami had the greatest impact on fisheries and aquaculture of any disaster, causing an estimated $500 million of damage in India and Indonesia. In Indonesia, the disaster almost paralyzed the fisheries sector and the livelihoods of communities that depend on it, with extensive damage to boats, harbours and fish ponds. Fisheries also suffer major impact in small islands developing states, because of the dependence of the SIDS on the fisheries sector, as well as the role the sector plays in food security and employment. In the Maldives, the sector was badly hit by the 2004 tsunami, with 70% of the economic impact on the agricultural sector associated with fisheries. Fishing harbours, boatsheds, fishing vessels and equipment, ocean cages, fish processors and equipment, fishery institutes and other assets were lost or seriously damaged (FAO, 2016).

The Sendai Framework for Disaster Risk Reduction 2015–2030 was adopted on 18 March 2015 with the goal to:

> *Prevent new and reduce existing disaster risk through the implementation of integrated and inclusive … measures that prevent and reduce hazard exposure and vulnerability to disaster, increase preparedness for response and recovery, and thus strengthen resilience.*

One of the main features of the Sendai Framework is a shift in focus from managing disasters to managing risks (UNISDR, 2015).

Four priorities for action were agreed:

Priority 1: Understanding disaster risk.
Priority 2: Strengthening disaster risk governance to manage disaster risk.
Priority 3: Investing in disaster risk reduction for resilience.
Priority 4: Enhancing disaster preparedness for effective response and to 'Build Back Better' in recovery, rehabilitation and reconstruction.

Reducing the risk of disaster by better understanding and assessing the risk, strengthening management processes and structures, and enhancing preparedness for a more effective response are priorities that fully align with actions aimed at adapting to climate change.

Adaptation and Mitigation Synergies

Although we have argued that mitigation should not be a priority for small island states, and that their efforts and investments should focus on adaptation measures, there are plenty of adaptation actions that produce mitigation co-benefits. Several of the 187 countries that produced communications on their Intended Nationally Determined Contributions (INDC) to climate change in advance of the COP21 conference in Paris in December 2015 noted that adaptation actions also generate mitigation co-benefits.

A total of 137 countries included an adaptation component in their INDC communications to the UNFCCC secretariat. Several countries indicated that adaptation is their main priority in addressing climate change (UNFCCC, 2016).

Table 3.13 summarizes the adaptation–mitigation synergies that were reported by the countries' communications to the UNFCCC secretariat (UNFCCC, 2016).

This is such a comprehensive list that it begs the question why any government should be focused solely on mitigation measures aimed at reducing emissions. Driving

Table 3.13 Adaptation and mitigation synergies.

Sector	Examples of adaptation measures with mitigation co-benefits
Agriculture, forestry, land use and livestock	• New crop varieties that require less use of pesticides and are water-stress tolerant • Sustainable land management practices • Improved livestock production practices • Protection and restoration of forests • Afforestation including mangroves and drought-tolerant species
Human settlements and infrastructure	• Climate-smart and resilient urban centres • Sustainable urban planning • Increased public transportation systems including electric vehicles • Waste management and treatment
Water	• Integrated water resources management, including watershed protection, wastewater and stormwater management, conservation, and recycling • Treated wastewater for crop irrigation • Wetlands restoration to promote absorption of carbon
Energy	• Transitioning to renewable energy – increasing the resilience of energy, water and health sectors • Energy efficiency
Ecosystems	• Marine park and coastal zone protection, blue carbon and seagrass beds • Combating desertification • Ecosystems-based adaptation
Tourism	• Ecotourism

a wedge between mitigation and adaptation, as if governments have to choose one or the other, is a false dichotomy. A much wiser policy would be to look for ways where you can do both.

The single most effective way that the world can bring down carbon emissions is for all countries to aggressively promote the transition to renewable energy for electrical power generation and renewable energy technologies for services such as lighting, water pumping, desalination, communication and transport. This is a transition that brings huge co-benefits in terms of mitigation and adaptation.

References

Alongi, D.M. (2008) Mangrove forests: Resilience, protection from tsunamis, and responses to global climate change. *Estuarine, Coastal and Shelf Science*, 76, 1–13.

Beebe, W. (1928) *Beneath Tropical Seas: A Record of Diving among the Coral Reefs of Haiti*. NY Zoological Society, Putnam and Sons, New York.

Bouman, B.A.M., Lampayan, R.M. and Tuong, T.P. (2007) *Water Management in Irrigated Rice: Coping with Water Scarcity*. International Rice Research Institute, Los Baños, Philippines.

Bryan, E., Deressa, T.T., Gbetibouo, G.A. and Ringler, C. (2009) Adaptation to climate change in Ethiopia and South Africa: options and constraints. *Environmental Science and Policy*, 12, 413–426.

Bush, S. (2014) *Natural resources management and ecosystem-based adaptation*. Report to the UNDP, Port-au-Prince, Haiti.

CBD (2009) *Connecting Biodiversity and Climate Change Mitigation and Adaptation*. Report of the Second Ad Hoc Technical Expert Group on Biodiversity and Climate Change. Montreal, CBD Technical Series No. 41.

CGIAR (2014) *Evidence of Impact: Climate-smart Agriculture in Africa*. Consultative Group on International Agricultural Research, Program on Climate Change, Agriculture and Food Security, Copenhagen.

Chinvanno, S., Souvannalath, S., Lersupavithnapa, B., *et al.* (2008) Strategies for managing climate risks in the lower Mekong River Basin: A place-based approach, in *Climate Change and Adaptation* (eds N. Leary *et al.*). Earthscan, London, pp. 228–246.

Cinner, J.E., Huchery, C., MacNeil, M.A., *et al.* (2016) Bright spots among the world's coral reefs. *Nature*, 535, 416–419. doi:10.1038/nature 18607

Clements, R., Haggar, J., Quezada, A. and Torres, J. (2011) *Technologies for Climate Change Adaptation – Agricultural Sector*. UNEP Riso Centre on Energy, Climate and Sustainable Development, Roskilde, Denmark.

Colis, A., Ash, N. and Ikkala, N. (2009) *Ecosystem-based Adaptation: A Natural Response to Climate Change*. IUCN, Gland, Switzerland.

Costanza, R., Pérez-Maqueo, O., Luisa Martinez, M., *et al.* (2008) The value of coastal wetlands for hurricane protection. *AMBIO*, 37(4), 241–248.

Dasgupta, S., Laplante, B., Meisner, C., *et al.* (2007) *The Impact of Sea level Rise on Developing Countries: A Comparative Analysis*. World Bank Policy Research Working Paper 4136. Development Research Group, World Bank, Washington, DC.

DeAngelo, B., Rose, S., Beach, R.H., *et al.* (2005) Estimates of joint soil carbon, methane, and N_2O marginal mitigation costs from world agriculture. In *Proceedings of the 4th International Symposium NCGG-4* (ed. A. van Amstel). Millpress, Rotterdam.

Delgado, J.A., Nearing, M.A. and Rice, C.W. (2013) Conservation practices for climate change adaptation. *Advances in Agronomy*, 121, 47–115.

Elliot, M., Armstrong, A., Lobuglio, J. and Bartram, J. (2011) *Technologies for Climate Change Adaptation – the Water Sector* (ed. T. de Lopez). UNEP Risoe Centre, Roskilde.

EPA (2004) *Guidelines for Water Reuse*. Report EPA/625/R-04/108. Environmental Protection Agency, Washington, DC.

FAO (2008) Rice and climate change. Available from www.fao.org/fileadmin/templates/agphome/documents/Rice/rice_fact_sheet.pdf

FAO (2010a) *Climate-smart Agriculture Policies, Practices and Financing for Food Security, Adaptation and Mitigation*. FAO, Rome. Available at www.fao.org/docrep/013/i1881e/i1881e00.pdf

FAO (2010b) *The wealth of waste: The economics of wastewater use in agriculture*. FAO Water Report No. 35. FAO, Rome.

FAO (2015) Economic aspects of conservation agriculture. Available from www.fao.org/ag/ca/5.html

FAO (2016) The state of the world fisheries and aquaculture. FAO, Rome. Available from http://www.fao.org/3/a-i5555e.pdf

FAOSTAT (2016) Food and agricultural data. Available from www.fao.org/faostat/en/#home

Gilman, E.L., Ellison, J., Duke, N.C. and Field, C. (2008) Threat to mangroves from climate change and adaptation options: a review. *Aquatic Botany*, 89, 237–250.

GWP (2014) *Integrated water resources management in the Caribbean: The challenge facing small island developing states.* Global Water Partnership. Available at http://infoagro.net/archivos_Infoagro/Ambiente/biblioteca/EN_04Caribbean_TFP_2014.pdf

GWP (2016) Rainwater harvesting in the Caribbean. Available from www.caribbeanrainwaterharvestingtoolbox.com/about.htm

IUCN (2014) What are protected areas? Available at www.worldparkscongress.org/about/what_are_protected_areas.html

IUCN (2016) *Explaining Ocean Warming: Causes, Scale, Effects and Consequences.* International Union for the Conservation of Nature, Gland, Switzerland.

Karnauskas, K.B., Donnelly, J.P. and Anchukaitis, K.J. (2016) Future freshwater stress for island populations. *Nature Climate Change*, 6, 720–725. doi:10.1038/nclimate2987

Kawasaki, J. (2010) Thailand's rice farmers adapt to climate change. United Nations University. Available from http://ourworld.unu.edu/en/climate-change-adaptation-for-thailands-rice-farmers

Kirshen, P., Knee, K. and Ruth, M. (2008) Climate change and coastal flooding in Metro Boston: impacts and adaptation strategies. *Climatic Change.* doi:10.1007/s10584-008-9398-9

Leip, A. and Bocchi, S. (2007) Contribution of rice production to greenhouse gas emissions in Europe. In *Proceedings of the 4th Temperate Rice Conference*, Novara, Italy.

Linham, M.M. and Nicholls, R.J. (2010) *Technologies for climate change adaptation – Coastal erosion and flooding.* UNEP Riso Centre on Energy, Climate and Sustainable Development, Roskilde, Denmark. Available at www.uneprisoe.org

Long, S.P., Ainsworth, E.A., Leakey, A.D.B., *et al.* (2006) Food for thought: Lower-than-expected crop yield stimulation with rising CO_2 concentrations. *Science*, 312, 1918–1921. doi:10.1126/science.1114722

Mainuddin, M., Kirby, M. and Hoanh, C.T. (2012) Impact of climate change on rainfed rice and options for adaptation in the lower Mekong Basin. *Natural Hazards*, 66, 905–938. doi:10.1007/s11069-012-0526-5

Nearing, M.A., Jetten, V., Baffaut, C., *et al.* (2005) Modeling response of soil erosion and runoff changes in precipitation and cover. *Catena*, 61, 131–154.

Nelson, G.C., Rosegrant, M.W., Koo, J., *et al.* (2009) *Climate Change: Impact on Agriculture and Costs of Adaptation.* International Food Policy Research Institute, Washington, DC.

Ngigi, S.N. (2009) *Climate change adaptation strategies: Water resources management options for small holder farming systems in Sub-Saharan Africa.* MDG Centre for East and Southern Africa. The Earth Institute of Columbia University, New York.

PA (2016) *Rainwater harvesting.* Practical Action, The Schumacher Centre, UK. Available at answers.practicalaction.org/our-resources/collection/rainwater-harvestin-1.

Padgham, J. (2009) *Agricultural development under a changing climate: Opportunities and challenges for adaptation.* Joint Departmental Discussion Paper, Issue 1. World Bank, Washington, DC.

PAI (2011) *Why Population Matters to Water Resources.* Population Action International, Washington, DC.

Peng, S., Huang, J., Sheehy, J.E., *et al.* (2004) Rice yield decline with higher night temperature from global warming. *Proceedings of the National Academy of Sciences*, 101(27), 9971–9975. doi:10.1-73/pnas.0403720101

Pinheiro, B. da S., de Castro, E. da M. and Guimarães, C.M. (2006) Sustainability and profitability of aerobic rice production in Brazil. *Field Crops Research*, 97, 34–42.

Porter, J.R., Xie, L., Challinor, A.J., *et al.* (2014) Food security and food production systems, in *Climate Change 2014: Impacts, Adaptation, and Vulnerability. Part A: Global and Sectoral Aspects. Contribution of Working Group II to the Fifth Assessment Report of the Intergovernmental Panel on Climate Change*. Cambridge University Press, Cambridge, pp. 485–533.

Reid, H. and Huq. S. (2005) Climate change biodiversity and livelihood impacts, in *Tropical Forests and Adaptation to Climate Change: In Search of Synergies* (eds C. Robledo, M. Kanninen and L. Pedroni). CIFOR, Bongor.

Reij, C., Tappan, G. and Smale, M. (2009) *Agroenvironmental Transformation in the Sahel: Another Kind of Green Revolution*. IFPRI Discussion Paper, International Food Policy Research Institute, Washington, DC.

Robbins, M. (2004) *Carbon Trading, Agriculture and Poverty*. World Association of Soil and Water Conservation (WASWC), Bangkok.

Salm, *et al.* (2000) *Marine and Coastal Protected Areas: A Guide for Planners and Managers* (3rd edn). International Union for the Conservation of Nature and Natural Resources, Gland, Switzerland.

Spalding, M., McIvor, A.,Tonneijck, F.H., *et al.* (2014) *Mangroves for Coastal Defence. Guidelines for Coastal Managers and Policy Makers*. Wetlands International, Wageningen, Netherlands and The Nature Conservancy, Arlington, Virginia.

Thompson, T., Kennedy, N., Thomason, W., *et al.* (2014) *LTRA-06 Update, 2014*. Sustainable agriculture and natural resources management (SANREM) project, Haiti. Virginia Tech.

Turral, H., Burke, J. and Faurès, J-M. (2011) *Climate Change, Water and Food Security*. FAO Water Report No. 36. FAO, Rome.

UNDESA (2014) *Trends in Sustainable Development: Small Island Developing States (SIDS)*. United Nations Department of Economic and Social Affairs. New York.

UNEP (2012) *Integrated Water Resources Management Planning Approach for Small Island Developing States*. United Nations Environment Programme, Nairobi.

UNFCCC (2016) *Aggregate effect of the intended nationally determined contributions: an update*. Synthesis report by the secretariat. FCCC/CP/2016/2. Available at http://unfccc. int/resource/docs/2016/cop22/eng/02.pdf

UNISDR (2015) Chart of the Sendai Framework for Disaster Risk Reduction 2015–2030. Available at www.unisdr.org/files/44983_sendaiframeworksimplifiedchart.pdf; www. unisdr.org/files/43291_sendaiframeworkfordrren.pdf

WHO (2006) *Guidelines for the Safe Use of Wastewater, Excreta and Greywater, Vol. 2: Wastewater Use in Agriculture*. World Health Organization, Geneva.

4

Adapting Energy Systems

The transition from fossil fuels towards renewable energy is generally characterized as a mitigation measure, since it is clear that replacing electrical power generation technology powered by oil, gas and coal, with zero-emission renewable energy will substantially reduce emissions of carbon dioxide and methane. Renewable energy technologies such as solar photovoltaic, solar parabolic systems, power towers, wind power, geothermal energy, and relative newcomers such as wave and tidal power, have the potential to replace practically all the fossil fuel generating systems operating around the globe. Whether a part of baseload power generation should be provided by nuclear energy is still being debated, but even without nuclear power there are many analyses that show that renewable energy can reliably provide enough electrical power, including baseload power, for all countries.

This is of course a massive transitional shift in technology akin to replacing horse-drawn carriages in the nineteenth century with the automobile. That transition in North America and Europe took several decades, and we are looking at a global energy transition of equivalent timescale.

The transition to renewable energy systems is an important adaptation mechanism for small island states because it substantially increases their ability to withstand energy price shocks and fuel supply disruptions. Recall that an important component of the vulnerability of the SIDS is due to their geographical isolation and their dependence on imported goods, including oil and gas for the generation of electricity. Climate change will increase the risks and the costs associated with this isolation. Although hydrocarbon fuels for automobiles are going to be necessary for at least a couple of decades (it is hard to see a widespread transition to electric vehicles in developing countries happening soon), the level of imported hydrocarbons can be substantially reduced if electrical power generation systems shift over to renewable energy (THEnergy, 2016).

Renewable energy is important for adaptation as well as mitigation because:

- the source of energy is national, dependable, almost uninterruptible, and free;
- renewable energy reduces expensive fossil fuel imports, improves the balance of payments position, and frees up foreign exchange that can be used for other climate-proofing action;
- electricity from renewable energy sources is substantially cheaper than power generated from imported petroleum fuels, so if these savings are passed on to households, renewable energy can be an important factor in alleviating poverty;

Climate Change Adaptation in Small Island Developing States, First Edition. Martin J. Bush.
© 2018 John Wiley & Sons Ltd. Published 2018 by John Wiley & Sons Ltd.

- distributed sources of power generation are inherently more resilient to extreme weather;
- PV systems on well-built infrastructure such as hospitals and schools serve as community protection centres in times of extreme weather;
- PV powered communication systems are better able to withstand storms and to continue to provide essential communication services;
- coastal installations for importing and stocking petroleum fuels for power generation are inherently vulnerable to coastal flooding and storm surge, so they would become obsolete;
- rural electrification reduces poverty – an essential factor in building community resilience to climate change.

Renewable energy programmes should also go hand in hand with energy efficiency programmes. Why? Because energy should not be wasted any more than water should be wasted. Investments to improve energy efficiency are also intelligent adaptation approaches because, once again, if you are a small island developing state, it makes sense to be as self-reliant as possible. A 15% improvement in energy efficiency means the same service can be provided by a renewable energy system that is 15% smaller – in fact, more than 15% smaller because of avoided transmission and distribution losses. That translates into a substantial reduction in upfront capital and installation costs for the renewable energy system. Where conventional electrical power generating systems are still the norm, energy efficiency measures are very often the most cost-effective option; rather than ramping up power generation to respond to increasing demand, it is less expensive to invest in energy efficiency measures that actually reduce demand; thereby allowing the same generating capacity to service more customers or cover a wider area.

Depending on Energy

The majority of the small island developing states are almost entirely dependent on imported petroleum products to generate electricity and to provide gasoline and diesel fuel for the transport sector. Only eight small island states possess petroleum resources that they are currently exploiting. Five are in the Caribbean: Trinidad and Tobago, Barbados, Cuba, Belize and Suriname. The other three oil-producing SIDS are Bahrain, Timor-Leste and Papua New Guinea. The level of oil production of Barbados and Belize is small: less than 2000 barrels a day (bbl/day). Oil production in the other countries in the group varies from 15,000 bbl/day in Suriname to 81,000 bbl/day in Trinidad (WFB, 2015). Most of these SIDS also produce natural gas – an efficient fuel for electrical power production.

For the 43 small island states without oil resources, the financial burden of importing substantial quantities of refined petroleum products is considerable. Only two states – Fiji and Belize – generate most of their power from renewable energy resources including hydropower. For 30 of the island states, their dependence on imported petroleum fuels for power generation is greater than 90%. Table 4.1 lists the SIDS that depend almost entirely on imported fuels for electrical power generation. The data are from 2012.

Table 4.1 Dependency of SIDS on imported petroleum fuels for electrical power generation (2012).

90–100%		80–90%	70–80%
American Samoa	Aruba	Dominican Republic	Cabo Verde
Antigua & Barbuda	Bahrain	St Vincent & Grenadines	French Polynesia
Bahamas	Comoros	Vanuatu	Haiti
Barbados	Cook Islands		New Caledonia
British Virgin Islands	Cuba		Samoa
Guam	Grenada		Sao Tome & Principe
Guinea-Bissau	Guyana		
Kiribati	Jamaica		
Maldives	Mauritius		
Monserrat	Micronesia Federated		
Nauru	Puerto Rico		
Niue	Singapore		
Seychelles	Saint Kitts & Nevis		
Saint Lucia	Tuvalu		
Tonga			
US Virgin Islands			

Source: WFB (2015).

Table 4.2 Five SIDS generating more than a third of their electricity from renewable energy sources (percentages).

	Petroleum fuels	Hydropower	Solar and wind
Belize	47	27	26
Dominica	60	18	22
Fiji	47	48	5
Papua New Guinea	61	31	8
Suriname	54	46	0

Source: WFB (2015).

Only five SIDS have less than a 70% dependency: Belize, Dominica, Fiji, Papua New Guinea and Suriname. These five states have developed and installed significant levels of renewable energy power generation, as shown in Table 4.2 (WFB, 2015).

Access to Electricity

Electricity is an essential form of energy for households and communities. Without electricity, opportunities for commerce are difficult, and for many families, developing a small business is impossible. Lighting a home with kerosene lamps is inadequate for

Table 4.3 SIDS with less than 90% access to electricity (2012).

Access to electricity (% of households)		
American Samoa 59	Guyana 80	Papua New Guinea 18
Cabo Verde 71	Haiti 38	Sao Tome & Principe 61
Comoros 69	Kiribati 59	Solomon Islands 23
Fiji 59	Marshall Islands 59	St Vincent & Grenadines 76
French Polynesia 59	Micronesia Fed. States 59	Timor-Leste 42
Guam 59	New Caledonia 59	Tuvalu 45
Guinea Bissau 61	Palau 59	Vanuatu 27

Source: World Bank (2016).

school children trying to study in the evening. Without electricity, rural families cannot communicate reliably with meteorological and disaster management agencies. They lack timely and accurate information about approaching storms and extreme weather. People cannot always charge their cellphones, and batteries for radios quickly go dead. Households without electricity are therefore especially vulnerable to rapidly evolving events such as violent storms, flooding, landslides and mudslides.

Adaptation to climate change aims to increase resilience and reduce vulnerability. Ensuring that at least nine out of ten households have access to electricity should be a priority for all the small island states.

For the SIDS, Table 4.3 shows the islands where less than 90% of households have access to electricity (World Bank, 2016). No data are available for Anguilla, British Virgin Islands, Cook Islands, Montserrat and Nauru.

Renewable Energy

The year 2015 was notable in that, for the first time, investments in renewable energy in developing countries (excluding large hydropower) exceeded those in developed countries. The developing world, including China, India and Brazil, committed a total of $156 billion to renewable energy, exceeding the amount invested by developed countries by $26 billion. Together, a record $286 billion was invested in renewable energy (excluding large hydropower) in 2015 – more than double the investment in new coal and gas generation (FS, 2016). Besides the three countries cited above, South Africa, Mexico, Chile, Morocco, Uruguay, the Philippines, Pakistan and Honduras each registered substantial investments in renewable energy. Figure 4.1 shows investments in renewable energy in developed and developing countries.

The amount of generating capacity installed as a result of this investment also broke records. For the first time, more renewable energy capacity was installed than capacity for conventional fossil fuel power systems.

In recent years, renewable energy has become increasingly dominated by wind and solar, with the smaller technologies losing ground, and in 2015 this trend became even more apparent (Figure 4.2). Photovoltaic energy and wind power led the way with a total of 118 GW installed in 2015 – a substantial increase over the 2014 figure of 94 GW.

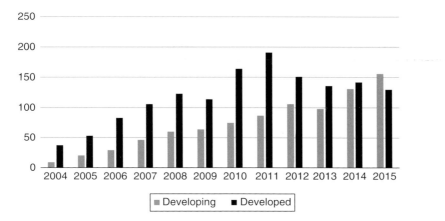

Figure 4.1 Investments in renewable energy technologies (billion $). Adapted from FS (2016). Reproduced with permission of Frankfurt School of Finance & Management gGmbH 2016.

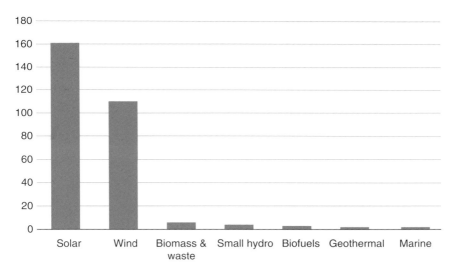

Figure 4.2 Global renewable energy investments in 2015 (billion $). Adapted from FS (2016). Reproduced with permission of Frankfurt School of Finance & Management gGmbH 2016.

Wind energy comes in two forms: onshore and offshore. Investment in onshore wind was only slightly higher in 2015 at almost $84 billion, while offshore wind, the smaller of the two, attracted a record $23 billion – up 39% compared with 2014. More than 20 offshore wind farms were financed worldwide in 2015. This growth was mostly in Europe, but there were new installations in China (FS, 2016).

Part of the reason for the dominance of wind and solar is that costs for PV energy and wind power installations are becoming increasingly competitive as costs continue to fall. Over the last five years the levelized cost of electricity from wind and solar has dropped substantially, as shown in Table 4.4.

In the second half of 2015, the global average levelized cost of electricity from crystalline silicon PV panels was $122 per MWh, down from $143/MWh a year earlier.

Table 4.4 Global average levelized cost of electricity for wind and solar.

RE technology	Levelized cost, Q3 2011, $/MWh	Levelized cost, Q4 2015, $/MWh
Onshore wind	95	83
Offshore wind	220	174
Solar PV crystalline silicon	200	122
Parabolic trough solar with storage	275	275

Source: Adapted from FS (2016).

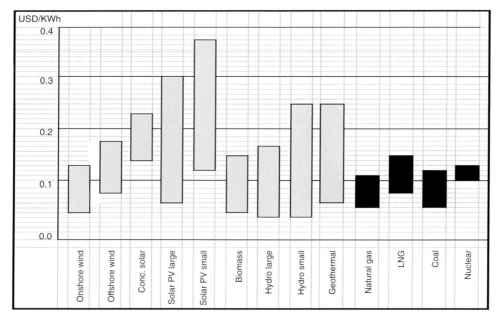

Figure 4.3 Levelized costs of electricity from renewable energy technologies in 2014. *Source:* Reproduced with permission of IRENA (2015).

However, several contracts signed in 2016 have come in substantially below these numbers: for example, bids in May 2016 for the megawatt-scale Mohammed bin Rashid Al Maktoun solar park being built in Dubai were reportedly as low as $29.9/MWh, equivalent to 2.99 ¢/kWh (EC, 2016).

Figure 4.3 shows the range of levelized costs of electricity from renewable energy technologies compared with fossil fuels and nuclear energy in 2014 (IRENA, 2015).

These costs do not include any externalities: the environmental costs of emissions from coal and natural gas, and the substantial costs of decommissioning nuclear plants. Not shown in this figure is the cost of generating electricity from diesel generators, which is anywhere from 0.35 to over 0.60 $/kWh.

The cost of electricity paid by residential consumers varies widely across the SIDS. One of the cheapest tariffs is perhaps that charged in Trinidad and Tobago: the national

Table 4.5 Electricity tariffs in Jamaica in 2011.

Category	US cents/kWh
Street lighting	41.7
Small commercial users	38.8
Residential customers	37.6
Large commercial customers	33.2
Industrial users	31.0
Other customers	29.0

Source: Adapted from Shakuntala *et al.* (2013).

power company, T&TEC, sells electricity to residential customers at around 4 cents/kWh. At the other end of the scale are countries where most of their electrical power is generated from imported petroleum fuels, and to cover those costs, electricity tariffs can be set as high as 50 cents/kWh. Among the Caribbean SIDS, the highest tariff in 2011/2012 reportedly was 33 cents/kWh, charged in St Lucia (Ochs *et al.*, 2015). However, a report in 2013 showed even higher tariffs for Jamaica, as shown in Table 4.5.

The high electricity tariffs being charged by many SIDS provide a considerable financial and economic incentive for governments to invest in less expensive renewable energy technologies that generate electrical power.

For the majority of the SIDS, the three main renewable energy technologies of photovoltaics, wind power and solar thermal energy have huge potential to:

1) break the costly dependency on imported petroleum fuels;
2) provide electricity to rural communities not yet connected to the grid; and
3) catalyze and drive the transition from fossil fuels to renewable energy.

We will look at each of these technologies in turn. At the end of the chapter we will review geothermal energy, hydropower, concentrating solar power, ocean thermal power and wave power, and explain why these technologies take second place – at least for the moment.

Photovoltaic (PV) Electricity

This technology is transforming the way the world produces electricity. It is not just that PV power is now no more expensive than electricity generated by conventional power stations burning coal and natural gas, and much less expensive than nuclear energy when all the external costs are factored in; it is the fact that PV systems are the ultimate in flexibility. A small 20-watt PV panel in Haiti will charge a cell phone and a LED light for the evening; an array of 30 100-watt panels on a rural clinic in Papua New Guinea will keep the lights on and the antibiotics at the right temperature in a DC-powered refrigerator; a large PV array on a shopping mall in Trinidad will generate hundreds of kilowatts and provide enough power for all the stores and the kids plugging in their Ipads; and megawatt-scale grid-connected PV systems covering hectares of land can provide power to thousands of homes and businesses. Not enough space for that? PV systems can be floated on reservoirs – a large floating system in England has 23,000 floating panels and generates 6.3 MW. This solar farm will soon be overtaken by a 13.7 MW floating PV array under construction (in 2016) in Japan.

Here we look at two of the most useful and cost-effective applications of PV power for small islands:

- Megawatt scale grid-connected systems
- Minigrid systems for rural areas

Like many small island states that are heavily dependent on imported fuel for power generation, Jamaica is planning a major programme to transition to renewable energy. In most of the Caribbean islands, that means photovoltaic energy and wind power. Jamaica has good solar and wind resources and is planning to exploit both sources of energy.

One of the largest PV systems in the Caribbean is installed on the roof of the Grand Palladium Resort Hotel on the northwest coast in Hanover, a county in the north of Jamaica. Operational since 2014, the 1.6 MW system is installed on the flat roofs of several of the hotel buildings. The system has 6300 silicon PV panels. The resort has a peak load of several megawatts, so the PV installation, large as it is, only provides about 10% of the electricity used by the hotel. The remaining demand is provided by the national grid. The grid-tied inverters are bi-directional, allowing for the hotel to provide power to the grid when the PV system is generating more than the hotel requires. A Power Purchase Agreement (PPA) with the Jamaican Public Service Company (JPS) sets out the different tariffs that apply.

The hotel invested $3.4 million in the system – just over $2 per peak watt, which represents a very good deal. A 1600 kW PV system should produce on average about 6400 kWh/day. The hotel is charged $0.30/kWh for electricity, so the savings add up to about $700,000 a year. At this rate, the hotel will recover its investment in less than five years.

The Caribbean is subject to fierce storms and hurricanes, so the PV panels are designed to withstand winds of 155 mph, or the equivalent of a Category 4 hurricane.

PV panels installed on flat roofs in the tropics have one additional and important benefit. They shade the roof surface, and if the roof has no thermal insulation – very often the case in the Caribbean – and if a significant fraction of the power is used for air conditioning, the PV panels will also reduce the air-conditioning load.

A last point concerns the Jamaican national grid. It is reportedly not in good shape, and so adding MW-scale renewable energy systems is not always going to be feasible without significant upgrades to the transmission system. In coastal areas far from the capital and where tourism infrastructure imposes a heavy demand for power, it makes sense to install independent solar and wind power systems that satisfy the local demand but which are not necessarily connected to the national grid.

A much larger PV system began operations in Jamaica in 2016. Developed by WRB Enterprises, a Florida-based firm, the 20 MW grid-connected facility in Content Village, Clarendon, has 97,000 photovoltaic panels and covers an area of 62 hectares (154 acres).

The PPA contract stipulates that JPS will buy the plant's electrical energy for the next 20 years at a price of $0.18/kWh. It is a fixed-price PPA with no escalation clause, so effectively the electricity becomes less expensive for JPS in real terms over the life of the system. Electricity in Jamaica is expensive, and is blamed for many of the problems confronted by Jamaican manufacturing industries. They complain that the high cost of electricity makes many of their products uneconomic compared with similar products manufactured elsewhere in the Caribbean – in Trinidad, for instance, where electricity costs less than 5 cents a kWh. If solar and wind energy can reduce the cost of electricity for the commercial and industrial sectors, this is another significant benefit resulting from the transition to renewable energy.

In the US Virgin Islands, Toshiba International Corporation supervised the construction of a 4 MW PV plant on Estate Spanish Town on the island of St Croix. Completed in 2014, the plant has close to 20,000 255-Wp PV panels connected to the Water and Power Authority (WAPA) Midland Substation through eight 500 kW inverters. At the same time, WAPA entered into a 20-year PPA with Toshiba for the integration of the 4 MW plant into St Croix's power grid. WAPA also signed PPAs with SunEdison for 25 years and Lanco Virgin Islands for 20 years for a total of 18 MW of peak capacity – approximately 15% of the Virgin Islands' peak load. The 20-year PPA with Toshiba International was set at $0.155/kWh, increasing annually by 1.5% up to $0.172/kWh over the lifetime of the project. This figure, although relatively high, is reportedly still less than WAPA's avoided cost of power generation from imported fossil fuels.

These two examples highlight the clear economic benefits of building MW-scale photovoltaic systems on islands that are generating electrical power using imported petroleum fuels, and therefore charging commercial customers and households high prices for electrical energy. In almost all cases, power purchase agreements (PPAs) can be negotiated with developers for a kWh price substantially less than the avoided cost of a kWh of energy generated from imported fuel.

Minigrid Systems

On the larger islands, many villages lack access to electricity because they are far from the national grid, and extending the distribution system to these communities is generally not economic for national power companies that need to cover their costs.

On many small islands, rural communities are supplied with electricity using diesel generators. Generating electricity in this way is expensive and unreliable: kW-scale generators need regular maintenance and transporting diesel fuel to rural communities is often difficult and expensive. In some cases, stand-alone wind or PV systems have been installed, but these systems can also be unreliable because of the intermittent nature of wind and the lack of solar energy at night. Battery storage mitigates these effects, but batteries are expensive and, if not carefully maintained (and kept cool), need to be replaced after only a few years of service. Combining both technologies makes it possible to offset the limitations of each of these technologies.

A PV hybrid system combines PV panels and a diesel generator to supply power to a rural minigrid. A bank of batteries is generally installed as this stores electricity when the PV panels generate excess energy, and directs this energy to the minigrid when the demand for electricity exceeds the power being generated by the PV array. The diesel generator provides the backup power necessary when the solar energy is insufficient and the batteries are discharged.

Figure 4.4 shows schematically the structure and configuration of the main elements of a PV-diesel hybrid system. The PV panels produce DC electricity which is converted

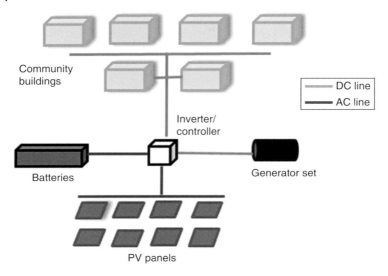

Figure 4.4 Schematic outline of PV-diesel hybrid system for rural electrification.

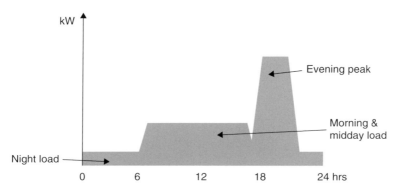

Figure 4.5 Typical load profile for a rural community.

to AC power by an inverter that also controls the system and manages the distribution of power: directing excess DC power to the battery bank and starting up the generator when the PV output and the batteries cannot meet the demand for electricity (IEA, 2013).

The typical load curve for a rural village is composed of a prominent peak in the evening as lights get switched on, a smaller peak at midday, and a much smaller baseload that remains more or less constant outside of the peak periods, including at night. A typical load profile for a village is shown in Figure 4.5.

If this electrical load were to be provided by a diesel generator sized for the peak load, it would run very inefficiently for most of the time, consuming much more fuel per kWh of electricity under the prevailing low load conditions. A PV hybrid system with batteries can supply the low load condition overnight from the batteries, while during the day the PV system provides power for the morning and midday loads. The diesel generator powers up to cover the evening peak and so runs only a few hours a day at its rated output.

Haiti has one of the lowest electricity connection rates in the world, at less than 40% of the population. Many of the households that have access to electricity are tapping into the distribution system illegally, which means that the national power company, Electricité de Haiti, never covers its costs and requires substantial support from the Government of Haiti in order to keep the country's inadequate and overloaded distribution system functioning for at least part of the time. Few areas in Haiti, even in the capital Port-au-Prince, have reliable electrical power. Families that can afford it will install a bank of 4 or 8 deep-cycle 6-volt batteries that are charged by an inverter when electricity is available, and which provide power to the house when the grid goes down. But an inverter-battery system is often insufficient – if the grid power goes off for several days for instance, a backup diesel generator is also required. This level of investment is of course impossible for the majority of Haitian families, and also for many businesses, so the lack of a reliable electricity supply system has a huge negative impact on commerce, services, and economic development in general.

Many rural areas in Haiti are far from any electricity distribution system, and so families have no choice but to rely on kerosene lamps for lighting. Kerosene is not cheap in rural areas, and so poor families are paying out a significant fraction of their income on inadequate and poor quality lighting.

In the town of Les Anglais in southwest Haiti, a PV-hybrid system provides power to homes and businesses. Set up in 2012 by EarthSpark International, a US-based non-profit organization, it is in many respects a conventional PV-hybrid system, but with several innovative features.

The Les Anglais minigrid serves over 2000 customers – both households and local businesses. A 93 kilowatt PV array generates electricity that is distributed by the minigrid installed by EarthSpark (see Figure 4.6). A bank of batteries provides 400 kWh of electrical storage, and a small 30 kW diesel generator provides backup power if the solar energy is inadequate. The system is designed to reliably provide electricity 24/7 to its customers. The backup generator is only used occasionally: over the 6-month period September 2015 to March 2016, it ran for a total of only 90 hours (EarthSpark, 2016).

One of the problems in managing a minigrid system in a rural area is that families newly connected to the electricity service may have difficulty managing the amount of electricity they consume, and then have to pay for. If they are unfamiliar with how electricity is used and charged, they may be presented with a bill they cannot pay. Disconnecting customers from the service because they cannot afford to pay for their electricity is not a solution over the longer term because the economics of the service improves as more paying customers connect to the grid. The smart meters installed by EarthSpark help solve this problem by requiring households to prepay for their power consumption. This approach is familiar to Haitian families accustomed to buying kerosene for lighting, charcoal for cooking, and top-up cards for their cell phone service. In effect, you pay in advance for what you need.

EarthSpark's smart electricity meters were developed to meet the special needs of minigrid operators in expanding access to electricity to new customers. Other innovative features include time-of-use pricing, which enables EarthSpark to charge a lower price for electricity during the middle of the day when energy costs are lower because power is drawn from the PV system. In addition, the consumption of power can be monitored across the grid at the household level. The data can be used to manage the load, to educate the community on energy efficiency, and to detect and evaluate

Figure 4.6 The inauguration of the Les Anglais photovoltaic hybrid minigrid system on 1 June 2015. Photo: EarthSpark, used with permission. (*See insert for color representation of the figure.*)

technical problems immediately they arise. EarthSpark plans to install 80 more minigrid PV-diesel hybrid systems in Haiti before 2020 (EarthSpark, 2016).

The Les Anglais system in Haiti is an excellent example of applying modern technology to a vexing development problem: how to bring electrical power to the rural areas of poor countries. Because of the development of increasingly affordable and reliable renewable energy technologies coupled with smart meters and control systems, it is no longer necessary to think in terms of extending transmission and distribution systems. PV-hybrid minigrids offer a practical and cost-effective alternative.

Unfortunately, Hurricane Matthew, a category 5 cyclone, ripped into southwest Haiti in early October 2016 and badly damaged the PV array. It is a lesson that needs to be learned as storms in the Caribbean become more intense: PV systems and wind power installations need to be able to withstand hurricane force winds. But building stronger PV array mountings and support structures does not solve the larger problem: nearly all the homes connected to the minigrid were destroyed by the storm.

Distributed PV Systems

Utility-scale wind and solar projects are conceptually similar to the conventional system of a large power plant providing electricity through power lines connected to thousands if not millions of homes. Small-scale systems take a different approach – involving thousands of people directly in the production of electricity for their own use. Even minigrids are similar to conventional systems, although obviously at a much smaller scale.

In 2015, over \$200 billion was invested in new renewable power generating capacity worldwide, and of this amount, over a quarter went towards projects of less than 1 MW – typically rooftop and small ground-mounted PV installations.

Falling costs and innovative financing mechanisms are putting small-scale distributed photovoltaic systems within the reach of people in both developed and emerging economies. Japan remains by far the largest small distributed power market, thanks to attractive feed-in tariffs (FITs) and falling PV system costs. In Australia, rooftop PV power is also booming: 1.4 million systems are now installed.

However, there is one aspect of distributed power systems that is of particular importance to small island states. Distributed PV systems have the potential to supply electricity during grid outages (including mini-grid failures) resulting from extreme weather, storms and floods. Distributed PV systems can significantly increase the resilience of the local electrical power system. To take advantage of this capability, the PV system must be designed with resilience in mind and combined with other technologies such as energy storage and backup generation. Electrical system resiliency focuses on:

- **Prevention** of power disruption
- **Protection** of life and property dependent on electrical power service
- **Mitigation** to limit the consequences of power disruption and failure
- **Response** to minimize the time it takes to restore service
- **Recovery** of electricity supply

Distributed PV systems may therefore play an important role in disaster management programmes. For instance, Florida's SunSmart Schools and Emergency Shelters Program has installed 115 10-kW PV systems with energy storage in schools to create emergency shelters. The programme provides teachers with opportunities for professional development and teaches students about alternative renewable energy and disaster preparedness (NREL, 2015).

The increased resilience of distributed PV systems means that they can be used to provide power to hospitals, emergency shelters, communications systems, and other infrastructure that would be incapacitated if the grid-connected power was disrupted during extreme weather.

In Haiti, where power outages are a routine occurrence, providing backup power to a hospital with diesel generators is not only expensive, but also unreliable. PV power is a cost-effective alternative. Healthcare centres consume lots of electricity. They operate around the clock, are often air-conditioned, and rely on lots of sophisticated equipment and machines, all of which require electricity. If the power goes off, patients' lives are at stake.

The PV-powered hospital in Mirebalais, Haiti, runs mostly on solar energy. The 300-bed hospital provides primary care to 185,000 people in Mirebalais and two neighbouring communities, also serving a much wider area with secondary and tertiary healthcare. It is also a teaching hospital.

The PV system consists of 1800 polycrystalline silicon panels rated at 280 W each, in total generating about 500 kW at maximum output. Five 95 kW grid-tied inverters produce AC power.

The Mirebalais hospital power system is in effect a hybrid system: combining grid electricity with solar PV power backed up by a diesel generator. Integrating three separate and independent power systems provides considerable reliability. An option would have been to install a bank of batteries – this would have reduced the use of the diesel generators but would have increased the cost of the system. The design of PV hybrid systems where power from the grid is unreliable and only intermittently available needs to carefully balance the different options in order to arrive at a least-cost design that provides the required level of reliability.

South Pacific island ditches fossil fuels to run entirely on solar power

A remote tropical island has catapulted itself headlong into the future by ditching diesel and powering all homes and businesses with solar energy. Using more than 5,000 solar panels and 60 Tesla power packs, the tiny island of Ta'u in American Samoa is now entirely self-sufficient for its electricity supply – though the process of converting to solar energy has been tough and plagued with delays.

Located 4,000 miles from the west coast of the United States, Ta'u has depended on over 100,000 gallons of diesel shipped in from the main island of Tutuila to survive, using it to power homes, government buildings and – crucially – water pumps.

When bad weather or rough seas prevented the ferry docking, which was often, the island came to a virtual stand-still. Utu Abe Malae, executive director of the American Samoa Power Authority, said Tutuila has subsidized Ta'u diesel shipments for decades to the tune of US$400,000 a year – and continually ran the risk of a serious environmental disaster if the delivery ships capsized during the notoriously treacherous journey.

Construction of the island's 1.4 megawatt micro-grid began two years ago, and was immediately bogged down by poor weather, transport delays and technical difficulties.

Solar engineers from contractors Tesla and SolarCity flew out from California to help oversee construction of the micro-grid, and 15 local men were employed in the construction process.

Five of the fifteen locals – previously low skilled, odd-job men on the island – have now transitioned to full-time jobs as solar power technicians managing the grid.

Source: The Guardian (2016).

Wind Power

In contrast to photovoltaic electricity – a technology of the twentieth century – the power of the wind has been harnessed by humans since the first fisher balancing in a dugout canoe had the idea of holding up and attaching a large piece of cloth and using the wind to propel his canoe forward. Fast-forward several millennia, and we now have massive 300-ton wind machines 100 metres tall capable of generating several megawatts of electrical power. These are sophisticated aerodynamic machines with little in common with the windmills of the past.

Wind power had another record-breaking year in 2015, as new installations topped 60 GW. At the end of 2015 about 433 GW of wind power was operating worldwide, and wind power supplied more new power generation that any other technology.

Onshore wind power generating costs have dropped significantly over the last few years. According to the International Energy Agency, in countries with good wind energy resources, onshore wind is now the cheapest electricity generation option, and costs are continuing to decline: contracted prices for onshore wind coming online in 2016–2017 are as low as 3 ¢/kWh (GWEC, 2016).

Two small island states are featured here as case studies: Samoa and Jamaica.

Samoa installed 500 kW of new wind power capacity in 2015 – its first wind power project. The turbines were erected on the island of Upolu, Samoa's second largest island. Two grid-connected 275 kW Vergnet turbines were installed by Masdar, a firm based in the United Arab Emirates with funding from the Pacific Partnership Fund (which is managed by the Abu Dhabi Development Fund) which has supported other renewable energy projects (mainly photovoltaic) in Pacific SIDS: Fiji, Kiribati, Vanuatu and Tuvalu.

The turbines on Upolu are expected to produce 1500 MWh of power annually, which would translate to a capacity factor of 31% – about right for wind turbines of this scale. This level of output reportedly saves about $475,000 in annual fuel cost savings on imported diesel fuel.

One interesting feature of the Vergnet turbines is that they can be lowered to the ground in the event that a cyclone is forecast to impact the island. The pylons are supported by cables that can, if necessary, be relaxed and extended in such a way that the column supporting the turbine and blades is lowered. This is obviously an important advantage for wind turbines installed in a region where extreme weather and more frequent cyclones are likely (Renew, 2016).

In Jamaica, the cost of importing petroleum products to fuel electrical power generation has had a substantial negative impact on the island's economy. The energy system is dominated by imported petroleum products which in 2009 accounted for 95% of the country's electricity generation. Consumers in Jamaica pay a high price for electricity (see Table 4.5), and for most customers, electricity prices more than doubled between 2005 and 2011 (Shakuntala *et al.*, 2013).

A 2011 survey of businesses in Jamaica found that high electricity costs were a leading cause of business failure. These results echo an earlier study by the Jamaica Manufacturing Association that found that the cost of electricity was the largest competitive disadvantage for local production when compared with regional competitors such as Costa Rica and Trinidad and Tobago. The cost of imported petroleum products is a huge burden on the economy. In 2011, imported fuel cost $2.2 billion (15% of GDP). Export revenues for the same year were estimated as $1.65 billion – not enough to cover the cost of imported fuel.

Under these economic circumstances, and with the cost of hydrocarbon fuels always volatile and unpredictable, Jamaica, like many other small island states, is aggressively moving towards renewable energy.

The potential for renewable energy in Jamaica, like the majority of small islands states that lie within the tropics, is considerable. A study by the WorldWatch Institute in 2013 noted the following key findings (Shakuntala *et al.*, 2013):

- Jamaica has excellent renewable energy potential especially for solar and wind: the entire island's electricity demand could be met with renewable resources.
- Several locations in Jamaica have extremely strong wind energy potential. Just 10 medium-sized wind farms could provide over half of the country's electrical power demand.
- Developing additional small hydropower capacity can provide cheap power to the electricity grid and energy access for remote locations.

Wind speed assessment began in Jamaica in 1995, which showed that sites along the southern coast had the highest wind speeds. As a rule of thumb, a good site for wind power will have a mean wind speed of at least 7 m/s. Based on these data, Wigton, with a mean wind speed of around 8 m/s, was selected for the installation of a 20 MW wind farm.

Wigton Windfarm Limited, a subsidiary of the Petroleum Corporation of Jamaica (PCJ), is the largest wind energy facility in the English-speaking Caribbean. Located in Rose Hill, Manchester, the wind farm comprises two plants: the 20.7 MW Wigton I, which began operating in 2004, and Wigton II, an 18 MW extension that was commissioned in 2010. Wigton III, which is scheduled to come on-stream in 2016, will bring the wind farm's total capacity to 62.7 MW.

The first phase of the Wigton wind farms saw the installation of 23 NEG Micon NM54 turbines rated at 900 kW, followed by Wigton II where 9 Vestas V80 turbines each rated at 2 MW were installed. The capacity factors of the turbines were reported as 35% and 33% respectively. These numbers indicate that the two wind farms should generate about 115 GWh of electrical energy a year, which reportedly substitutes for about 60,000 barrels of imported oil (WorldWatch Institute, 2016).

The construction of the first wind farm was managed by RES Americas under contract to Wigton Wind Farm Ltd (WWF), a subsidiary of the Petroleum Corporation of Jamaica. Construction was swift; the first turbine was installed in mid-February 2004 and the wind farm was operational three months later.

The PPA with Jamaica Public Service Co. (JPS) for Wigton I was for energy at 5 ¢/kWh, a figure that was subsequently found to be too low for WWF to cover its operating costs. The PPA for Wigton II was reportedly set at 14 ¢/kWh, declining over the payback period to give a mean price of 10 ¢/kWh. This price is still below the avoided cost of JPS's power generation system based on imported fuel. For WWF, this price ensured that the wind farm business was profitable (WorldWatch Institute, 2016).

Cabo Verde is another island state with a major wind power programme. The 25.5 MW Cabeolica windfarm provides power to four islands in the archipelago. Situated 570 km off the west coast of Africa, Cabo Verde has excellent wind resources with many sites showing mean wind speeds in excess of 10 m/s. Costing a reported $85 million, the wind farm consists of 30 Vestas V52 turbines. If the capacity factor is about 30%, the wind farm should generate about 67,000 MWh per year.

Solar Water Heaters

The most efficient way to capture the energy of the sun is to absorb it onto a surface and find a way to convert the high surface temperatures that result into useful energy or work. One application that matches the incoming short-wave solar energy with the end-use application is solar water heating.

Heating water for domestic, residential and commercial use can consume a lot of electricity in places where natural gas is unavailable (the majority of the SIDS) and where electricity is used for heating water. This is not an efficient use of a high-quality energy form like electricity, but in many situations electricity is the only option. Along with air-conditioning and refrigeration, water heating using electricity is a major part of residential electricity consumption.

Table 4.6 Factors that facilitated market penetration of solar hot water heaters.

Action by	Target group	Factors contributing to commercial success
Private sector	Consumers	• High quality dependable product • Consumer guarantee • Finance to spread upfront cost of SWH unit • Community engagement and job creation • Strong marketing and communications
Private sector	Government	• Demonstrated potential of the technology • Cost-effective technology that saves consumers' money • Reduces household electricity consumption and therefore fossil fuel imports
Government	Consumers	• Involvement and participation through communications • Fiscal incentives (e.g. Homeowner Tax Benefit) • Increased duty on gas and electric water heaters
Government	Private sector	• Fiscal Incentives Act 1974 • Government purchase of SWH for new housing and buildings • Enabling environment of regulatory certainty and continued support

Source: CDKN (2016).

Two island states have pioneered the development of commercial-scale solar water heaters: Cyprus and Barbados.

The solar water heating business has been flourishing in Barbados since the 1970s, and over 50,000 solar water heaters have been installed. One estimate is that solar water heaters save over 100,000 MWh of electricity per year – more than the output of a 30 MW windfarm or roughly three times the output of a 25 MW photovoltaic system.

There are lessons to be learned from the experience in Barbados that apply in equal measure to the introduction of other distributed renewable energy systems – like household PV systems for instance. Table 4.6 summarizes the factors that led to success in Barbados.

Collaboration between the private sector and the government was essential. Without government support in the form of fiscal measures and tax breaks, it is unlikely that the construction and installation of solar water heaters would have been successful. At the same time, a good quality and reliable product that could be paid for over a period of time was also a key factor. Studies have shown that the cost of government incentives represented good value. In 2002, the tax incentives offered by the Government cost only about $500,000, while savings to households amounted to a minimum of $10 million annually (CDKN, 2016). In addition, baseload power consumption for electric water heaters would clearly have dropped by several megawatts.

The government of Barbados intervened strongly in support of the developing solar water heater market. Key interventions included:

• **1974 – Fiscal Incentives Act**. The Government introduced a tax exemption for the materials used to fabricate the solar water heaters, saving 20% of their cost. At the same time, a 30% tax was imposed on electric water heaters.

- **1977 – Government purchase of solar water heaters for state housing**. The Government mandated the installation of solar water heaters on new government-funded housing developments.
- **1980–82 – Homeowner tax benefits**. The Government made the cost of a SHW unit tax deductible up to a maximum of about $1750. This incentive was stopped in 1992 as part of economic restructuring following the recession in the late 1980s.
- **1996 – Amended homeowner tax benefit**. The homeowner tax benefit was reinstated in an amended form that allowed homeowners a tax deduction of $1750 for home improvements including mortgage interest, repairs, renovations, energy efficiency measures and the installation of solar water heaters (CDKN, 2016).

The long-term commitment of the Government of Barbados to the solar hot water heater programme is striking. It is doubtful the solar water heater market would have developed so strongly and so rapidly without this support.

It is worth noting that solar water heaters combined with PV electrical power can substantially reduce the energy demand of large hotels. Given the importance of the tourism sector in many small island economies and the difficulty many islands have in providing sufficient power from the main grid, small island governments should look closely at mandating these renewable energy systems in tourism infrastructure.

Hydropower

Hydropower has been the star renewable energy technology globally for more than a century. The sheer scale of the largest hydropower dams often challenges belief.

But there is a version of hydropower at a much lower scale that has a much smaller environmental and social footprint. It does not require a dam – just small-scale construction to divert water from a river into a pipe so that the water falls naturally (but under pressure) down to a turbine that generates electricity. Called run-of-the-river hydropower, it is a useful and generally cost-effective technology.

However, for many small island states, small-scale hydro is not necessarily the best option. Although the cost of small hydro compares favourably with other renewable energy technologies, the principal disadvantage is that the resource is only available in a few places on islands that have reliable rivers flowing down relatively steep terrain. The river does not need to be large, but its flow needs to be dependable. Several of the Caribbean SIDS have hydropower potential, including Haiti and Jamaica, but given the flexibility, modularity, ease of installation and decreasing cost of PV systems, solar energy and wind power are likely to be the preferred options.

Some of the larger SIDS (Guyana, PNG, Belize) might consider MW-scale hydropower schemes where dams are constructed across rivers and large quantities of water are impounded behind the dams. But the environmental and social impacts of large hydropower installations are exceptionally high, and recent research shows that greenhouse gas emissions from reservoir water surfaces are much greater than previously thought (Deemer *et al.*, 2016).

Geothermal Energy

A few islands have geothermal potential: Dominica, Montserrat, and the Grenadine islands in the Caribbean are volcanic islands with potential. But the technology is very capital-intensive and complex. In Djibouti, for example, where geothermal resources are reckoned to be exceptionally good, the technology has been studied for over 30 years. Test wells have been drilled, drilled again, then drilled elsewhere. The results are promising and the economics look good – on paper. But donors are wary of investing in a technology with high upfront costs, long lead times, complex engineering, and uncertain environmental impacts.

However, in 2015 the Green Climate Fund approved a programme called 'Sustainable Energy Facility for the Eastern Caribbean', which will support the development and exploitation of potential geothermal sites on Dominica, Grenada, Saint Kitts and Nevis, Saint Lucia, and Saint Vincent and the Grenadines. It remains to be seen whether geothermal energy at these sites will be possible, but given the high cost of electricity on these islands, there is a reasonable chance that electricity generation from geothermal energy will prove to be an economic proposition.

None of the other renewable energy technologies such as tidal power, ocean thermal energy conversion (OTEC) or wave power, are sufficiently well developed technically and commercially to be seriously considered by small island states. This situation may change, but at this time (2017) and for the foreseeable future, solar photovoltaic energy and wind power are by far the best options for small island developing states.

Solar Thermal Power

Although this technology is still not fully at scale, there is no question that in a few years time, concentrating solar power in a 'power tower' configuration may well be the solar technology of choice for megawatt-scale power generation. It has one huge and game-changing advantage over PV systems: it can generate electrical power 24 hours a day by using a thermal energy storage system, so that electricity can be generated during the night from heat generated and stored during the day. Small island states should continue to monitor the development of this technology. However, it works best in locations that have almost constant strong sunshine, since, unlike PV, it relies entirely on direct insolation. Not all the SIDS have solar energy resources equal to the task.

---

14 Pacific island nations negotiate world's first climate treaty to ban fossil fuels

As coastal erosion and sea-level rise eat away at the Solomon Islands due to climate change, the Pacific island nations are considering the world's first international treaty that would ban or phase out fossil fuels and set goals for renewables. The 'Pacific Climate Treaty' is currently under consideration after the fourth annual Pacific Islands Development Forum (PIDF) held in the Solomon Islands in July 2016.

During the two-day summit, 14 presidents, prime ministers and ambassadors from the island countries and territories discussed solutions to the Pacific's development challenges. The treaty is being utilized as a way to implement the aspirational 1.5 degrees Celsius target set by the Paris COP21 climate talks in the Pacific region, according to the Pacific Islands Climate Action Network (PICAN), a coalition of NGOs that wrote the treaty. The proposed treaty will be studied and a report will be presented at the 2017 summit.

The PIDF was created in 2013 by Fiji. This year's summit excluded Australia and New Zealand, which were part of earlier talks. At last year's talks, Australia and New Zealand were criticized by their smaller and developing island neighbours for having less ambitious climate change targets and for not doing more to combat climate change.

Many low-lying nations are under threat as oceans continue to rise. Scientists predict that Kiribati – a remote Island Republic in the Central Pacific – could be lost to rising sea-levels in the next 50 years.

The Philippines is particularly vulnerable to extreme weather, with the nation suffering violent storms like Typhoon Haiyan. Tropical storms have struck the nation more often and more severely, scientists believe, because of climate change.

PICAN said in a report presenting the Pacific Climate Treaty that the potential treaty parties 'already possess the political courage and commitment needed to adopt a flagship legal instrument that is sufficiently ambitious to prevent catastrophic changes in the global climate system ... Such a treaty, when implemented in collaboration with PIDF and civil society, would send a powerful signal to markets, governments and civil society around the world that the end of fossil fuels is near, with Pacific Islanders acting not as victims of climate change but as agents of change,' it said. 'As there is currently no treaty that bans or phases out fossil fuels, the Treaty would set a pioneering example to the rest of the world.'

The treaty includes sections on climate-related migration and adaptation. It would also set up a fund to compensate for communities that have suffered from climate change.

Source: Ecowatch (2016).

Energy Efficiency

Adaptation responses to climate change should also include measures to increase the efficiency of energy use. A detailed analysis of energy demand is essential, but the principal areas of focus are likely to be:

- the electricity distribution system
- the transport sector
- the building and construction sectors.

The distribution of electricity on many small island states is notoriously inefficient. Losses are due both to the poor condition of the distribution network and illegal connections that steal electricity from the system. Improving the distribution system should be a priority for small island governments – especially because the transmission and distribution systems are often under the management of a government agency that requires substantial subsidies in order to continue operations.

A first step is therefore the introduction of measures to increase revenue. These measures should include robust prevention of the theft of electricity. This is not difficult, because the illegal wires are easily detected from the street. What is required (apart from the willingness of the government agency to take on the job) is regular patrolling of the areas where illegal connections are commonplace by a mobile team of inspectors capable of handling any tension and conflict that may arise. Households should be warned not to continue the practice and, ideally, should be offered the possibility of connecting legally to the grid.

In low-income areas where the collection of revenue is difficult and inadequate, smart meters and prepaid metering (like the system installed in Les Anglais, Haiti) should be considered. The introduction of prepaid metering does not necessarily reduce the theft of electricity, so regular monitoring of neighbourhoods where theft is common will still be required.

The production and generation of electricity needs to be opened up to the private sector. Independent power producers (IPPs) need to be encouraged, and establishing the regulatory framework that enables this transition to take place should be a priority for governments. The law regulating Power Purchase Agreements (PPAs) needs to be articulated and made operational. If grid-tied residential PV systems are considered feasible (they require an adequate and reliable grid), then the technical specifications and the regulatory framework for these installations, and the feed-in-tariffs that apply, need to be legally established.

Many island governments have a system in place designed to ensure that private road vehicles are safe to use and roadworthy. In general, a government-operated licensing system requires that vehicles are inspected annually and, if approved, may be issued a license to operate. But in many countries this system is poorly regulated and routinely circumvented. Inspections need to be more detailed and more strictly applied. Public transport should obviously be a priority, and governments should carefully examine the possible savings to be made by introducing new, more energy-efficient buses onto the bus routes in urban areas, and on inter-city routes. In countries where the private sector operates pickups and minivans that function as taxis and public transport, these vehicles should be strictly inspected and regulated.

Buildings and other built infrastructure are generally major consumers of electricity used for applications such as water heating, air-conditioning, lighting and refrigeration. If electrical power is primarily generated from imported petroleum products, the demand for electricity from households, government buildings, hotels and the commercial sector should be seriously examined in order to identify and introduce ways to reduce the demand for electricity. There are many cost-effective options:

- Replacing electric water heaters with solar water heaters.
- Introducing more efficient air-conditioning equipment, such as split systems instead of window units, and central air-conditioning systems in larger buildings and tourism infrastructure.
- Photovoltaic systems on all new government buildings and a retrofit programme on older construction.
- Efficient light bulbs (LED and compact fluorescent), and banning the importation of incandescent light bulbs.
- Thermal insulation on all construction – mandated by legislation and revisions to the national Construction Code.

Governments need to intervene actively in order to incentivize these changes in behaviour and usage. Import duty should be lifted on all renewable energy and energy-efficient equipment (especially PV panels, wind turbines, power management and inverter units, smart meters, and LED lights), and materials used to build solar water heaters. Taxes should be imposed or increased on equipment that is inefficient: air-conditioning wall units, incandescent light bulbs (preferably banned completely), and electric water heaters.

References

CDKN (2016) Climate and Development Knowledge Network: Inside stories on climate compatible development. Available at http://cdkn.org/wp-content/uploads/2012/09/Barbados-InsideStory_WEB.pdf

Deemer, B.R., Harrison, J.A., Li, S., *et al.* (2016) Greenhouse gas emissions from reservoir water surfaces: a new global synthesis. *Bioscience*, 66 (11), 949–964. doi:10.1093/biosci/biw117

EarthSpark (2016) The evolution of a model: Pre-pay grid electricity in Haiti. Available at www.earthsparkinternational.org/blog/the-evolution-of-a-model-pre-pay-grid-electricity-in-haiti

EC (2016) New record set for world's cheapest solar. Available at http://www.ecowatch.com/2016/05/04/worlds-cheapest-solar

Ecowatch (2016) 16 July 2016. Available at www.ecowatch.com

FS (2016) *Global Trends in Renewable Energy Investment 2016*. Frankfurt School, FS-UNEP Collaborating Centre, Frankfurt, Germany.

GWEC (2016) *Global Wind Report: Annual Market Update 2015*. Global Wind Energy Council, Brussels.

IEA (2013) *Rural Electrification with PV Hybrid Systems: Overview and Recommendations for Further Deployment*. International Energy Agency Photovoltaic Power Systems Programme. Report IEA-PVPS T9-13:2013.

IRENA (2015) Rethinking energy: renewable energy and climate change. Available at www.irena.org/publications

NREL (2015) *Distributed Solar PV for Electricity System Resiliency*. National Renewable Energy Laboratory, US Department of Energy.

Ochs, A., *et al.* (2015) *Caribbean Sustainable Energy Roadmap and Strategy (C-SERMS): Baseline Report and Assessment*. Worldwatch Institute, Washington, DC.

Renew (2016) Samoa inaugurates 1st wind farm as Pacific turns away from diesel. Available at http://reneweconomy.com.au/2014/samoa-inaugurates-1st-wind-farm-as pacific-turns-away-from-diesel-71201

Shakuntala, M., Ochs, A., *et al.* (2013) *Jamaica Sustainable Energy Roadmap: Pathways to an Affordable, Reliable, Low-emission Electricity System*. Worldwatch Institute, Washington, DC.

The Guardian (2016) South Pacific island ditches fossil fuels to run entirely on solar power. *Guardian*, 3 December 2016.

THEnergy (2016) Database: Wind and solar power plants. Available at http://www.th-energy.net/english/platform-renewable-energy-on-islands/database-solar-wind-power-plants/

WFB (2015) *World Factbook*, available at https://www.cia.gov/library/publications/the-world-factbook/geos/aq.html. US Government, Central Intelligence Agency, Washington, DC.

World Bank (2016) Access to electricity (% of population). Available at data.worldbank.org/indicator/EG.ELC.ACCS.ZS

WorldWatch Institute (2016) Wigton Wind Farm: Jamaica's commitment to renewable energy starts paying off. Available at http://blogs.worldwatch.org/revolt/wigton-wind-farm-jamaicas-commitment-to-renewable-energy-starts-paying-off/

5

Managing Adaptation

The Key Climate Hazards

The 21st Conference of Parties to the United Nations Framework Convention on Climate Change (UNFCCC) that took place in Paris in December 2015 mobilized nearly all the countries of the world in an unprecedented effort to coordinate a global response to the threat of climate change. One hundred and eighty-nine countries submitted communications called Intended Nationally Determined Contributions (INDCs) which set out their plans to reduce emissions of greenhouse gases.

Of this number, 137 countries, including 46 developing countries, included an adaptation component in their INDC communications. Some of the countries indicated that adaptation was their main priority in addressing climate change (UNFCCC, 2016a).

It is instructive to look at how the governments of the countries that included an adaptation response in their INDCs see the dangers of climate change. Figure 5.1 shows the principal threats as identified by the countries themselves – the ordinate shows the number of countries identifying that threat as a key hazard.

The data suggest that most countries have correctly identified the key threats. But how should governments structure and marshal their resources in order to respond effectively to these threats?

Government ministries in developing countries tend to be compartmentalized along traditional lines. Apart from the core ministries – economy, finance, industry, education and health – there is almost always a Ministry of Agriculture (which may include fisheries). There is generally a Ministry of the Environment, although this may be a newer ministry. There may be a Ministry for Forestry, either separately or in combination with either Agriculture or Environment. A Ministry of Energy is unusual – energy is more likely to be a department in a ministry for infrastructure, transportation, communication or any combination of these sectors. The government agency that manages electrical power generation, transmission and distribution (generally a government-controlled monopoly) may be the only entity that actually concerns itself with electricity supply and demand. Another agency may regulate fuel prices for gasoline and diesel. Water is another sector that is often considered to be secondary and therefore managed as a department in a ministry considered to be more important. Coastal zone management may be under the Ministry of Environment, but this ministry is generally a lot less well financed then the principal ministries including Agriculture.

Climate Change Adaptation in Small Island Developing States, First Edition. Martin J. Bush.
© 2018 John Wiley & Sons Ltd. Published 2018 by John Wiley & Sons Ltd.

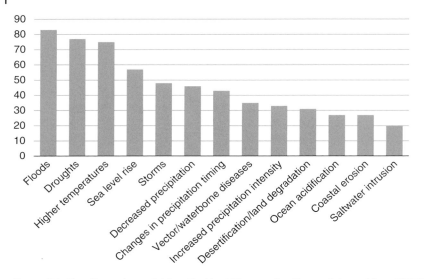

Figure 5.1 Key climate hazards identified by 137 countries. *Source:* Adapted from UNFCCC (2016a).

What generally does not exist is a ministry or a government agency for climate change – either mitigation or adaptation.

Ministries compete for money. The Ministry of Agriculture is generally one of the bigger ministries – quite often larger than the Ministry of Environment, and with more staff, vehicles and equipment. Agriculture may also have regional offices. For these reasons the Ministry of Agriculture is often funded much more generously than the Ministry of Environment. This is generally justifiable because in most developing countries, even those with a substantial tourism sector, agriculture still employs a large fraction of the population, including many of the poorest people. In addition, food security is a major concern for nearly all developing countries and the SIDS.

The question is: which government agency should take the lead in formulating and implementing the national action plans intended to reduce and manage the threats of climate change?

Figure 5.2 shows the priority areas and sectors for actions identified in the adaptation component of the communications on intended nationally determined contributions, where the ordinate level indicates the number of countries referring to an area or sector.

What is clear is that climate change adaptation is a cross-cutting theme that does not fit easily into the traditional silo-type structure of conventional government organization.

In order to manage climate change programmes and to coordinate appropriate action, there are generally three approaches:

1) Put the climate change agency under one of the line ministries – often this is the ministry of environment.
2) Create a high-level inter-ministerial committee that deals with the issues and formulates policy.
3) Create a new agency with the power to develop and manage the climate change programme.

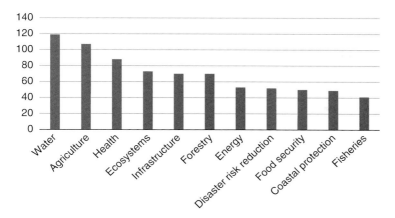

Figure 5.2 Priority sectors for adaptation. *Source:* UNFCCC (2016b).

The problem with the first option is that a Ministry of Environment very probably lacks both the resources and the technical competence to manage all the cross-cutting elements of climate change adaptation. Health and food security are not going to be areas where the Ministry of Environment has a great deal of experience or competence. Establishing and maintaining working relations with ministries not traditionally in partnership arrangements is often fraught with problems concerning leadership and accountability.

The second option is useful in that it permits the government to bring together experts from a range of ministries and agencies with the experience and competence to understand the issues and propose effective solutions. But an inter-ministerial committee lacks authority. It can formulate policy, draft action plans, hold workshops and organize consultative meetings, but its funding is probably limited and may only cover its operating costs. It may have no full-time staff except for a small secretariat. It is not the best option.

The third option is more difficult, takes longer to set up, and is more expensive. But it is the only option that establishes an effective and operational entity capable of dealing with the multiple threats of climate change. In the UK, for instance, in response to concerns about climate change, the government first set up the Office for Climate Change in 2006 with the mandate of coordinating government efforts to develop climate change policies and strategies. The objective was to improve intra-departmental communication and to coordinate and maximize climate change programmes across the different government units involved. Then, in 2008, Prime Minister Gordon Brown created the Department of Energy and Climate Change (DECC) with the aim of establishing an agency that could address the challenges of both climate change and energy supply.[1]

The government agency that is legally established to develop climate change policy, strategy and action plans should be operational at the highest government level, with the authority to participate and intervene in the development of the different ministries'

---

1 In July 2016, the Department of Energy and Climate Change became part of the Department for Business, Energy, and Industrial Strategy.

programmes of work and action plans. In small island states, this means that the climate change agency requires the backing and firm support of the office of the President. The climate change agency must be able to coordinate and collaborate with the principal ministries on an equal footing. The agency must have its own funding and budget; it will need fully-furnished office space, full-time technical and secretarial staff, and logistical resources such as vehicles and drivers. Staffing can be arranged by seconding many of the technical staff from the ministries and agencies that will be involved in climate change policy and planning: agriculture, energy, environment, water, electricity, health, forestry and fisheries. This arrangement facilitates coordination and inter-agency collaboration.

The national agency for climate change action, NACCA,[2] should *not* be an implementing agency. It leads the national programme on climate change mitigation and adaptation. It develops policy, defines sectoral strategic focus, coordinates action, manages the collaboration with funding agencies, oversees the programme, and evaluates and reports on results. It also ensures compliance with the UN conventions and protocols the country has signed up to and ratified.

As an example of the difficulty of coordinating action across government agencies, we can look at the ministries in Haiti responsible for actions that might be required to implement a programme of climate change adaptation. Four ministries share the important responsibility for the management of the environment and natural resources in Haiti:

- the Ministry of Environment (MDE);
- the Ministry of Agriculture, Natural Resources, and Rural Development (MARNDR);
- the Ministry of Planning and External Cooperation (MPCE);
- the Ministry of Public Works, Transport and Communication (MTPTC).

In the first national communication report on climate change, the responsibilities of these ministries were defined as given in Table 5.1 (MDE, 2001):

There is considerable overlap among the ministries, particularly concerning the management of natural resources and the environment. The two most important ministries are the Ministry of Environment (MDE), and the Ministry of Agriculture (MARNDR). Over the last few years, as climate change and its environmental impacts have come to the fore, the MDE has become much more involved in managing protected areas and key ecosystems. Not included above, however, is the Ministry of Interior and Territorial Collectivities (MICT), which is responsible for managing the Communes, and which oversees the important Department for Civil Protection (DCP), and the Inter-ministerial Committee for Land Use Management (CIAT), which has become an important actor on the environmental scene including climate change adaptation. In Haiti, there are therefore six agencies involved in climate change adaptation programmes.

Although the mandate of the MDE defines a stronger role than the MARNDR in terms of the protection of the environment including key watersheds, the MDE is much more centralized than the MARNDR, and has few staff based in the Departments or in

---

2 NACCA is the acronym employed in this chapter to refer to whatever national entity is established to tackle climate change.

Table 5.1 Ministries managing climate change action in Haiti.

Ministry	Directions, services, and associated agencies	Principal responsibilities
MDE	• Cabinet of the Minister • General direction • Technical direction • OSAMH	• Environmental policy and strategy • Promotion, management and conservation of forests, natural parks, management of buffer zones, legal and institutional framework • Environmental Action Plan • Protection of watersheds, soil and water conservation • Water policy; management of potable water; wastewater management
MARNDR	• Direction of Natural Resources • Parks Service (SPNS) • Service for water resources (SNRE) • Service for land restoration • Service for fish and aquaculture • Service for forestry resources	• Management of natural resources: soil, wood, vegetative cover, surface waters and aquifers, watersheds, meteorology and fishing
MPCE	• Direction for land management and protection of the environment	• Zoning of national territory; land use strategies; administrative division of territory; norms and national standards
MTPTC	• Office of Mines and Energy (BME) • EDH (Electricity of Haiti) • Direction urbanism • Direction sanitation and potable water (DINEPA)	• Research on exploitation of mineral and energy resources • Research and promotion of improved charcoal stoves • Hydropower • Urban and industrial wastewater management • Potable water system

Source: MDE (2001).

the Communes. In addition, the budget of the MDE is inadequate. The 2012–13 budget of the MDE was only US\$20.52 million; that of the MARNDR was ten times larger.

Although the MARNDR has a much larger budget, its presence in the Communes through the BAC offices (Bureau Communal de l'Agriculture) is weak and very often absent, particularly in the rural Communes. In sum, neither ministry has a strong presence in the watersheds and coastal areas likely to be included in a national adaptation programme.

Although the MARNDR and the MDE have agreed in the past to work together and to coordinate their activities, this objective has rarely, if ever, been achieved. As an example, a policy document prepared by the MARNDR in 2011 does not even mention climate change, and there is absolutely no reference to the Ministry of Environment (Bush and Sildor, 2009; MARNDR, 2011).

The Regulatory Framework

A substantial amount of legislative work will be required by the governments of small island States in order to establish the laws that permit, incentivize and regulate actions to promote government policy on climate change. These actions might include those under the following headings.

Environment

- Specifying pollution control measures and penalties for all activities negatively impacting coastal areas.
- Defining and enacting environmental law pertaining to the management and operation of major industrial installations: ports, refineries, cement production, factories, and so on.
- Defining and enforcing wastewater treatment regulations and setting construction norms for sewage treatment and disposal, especially for coastal infrastructure and all beach hotels.
- Specifying and enforcing coastal zone construction setbacks (the minimum distance from the high tide line where a building may be constructed).
- Zoning coastal areas and marine resources: mangroves, wetlands, estuaries, coral reefs and beaches.
- Establishing protected areas on all uninhabited islands.
- Establishing protected landscapes and zoning arrangements along the *entire* island coastline.
- Establishing marine protected areas, approving their management plans and zoning arrangements, and mandating adequate funding.
- Mandating the meteorology service to collect, analyze and report on solar insolation and wind speed data from priority island sites.

Energy

- Legalizing and defining the regulations for independent power producers (IPPs), including electricity from wind power and solar energy.
- Setting out the technical specifications for residential photovoltaic systems that are connected to the grid.
- Defining and regulating the terms of Power Purchase Agreements (PPAs) with IPPs.
- Defining energy efficiency standards for electrical equipment and appliances: air-conditioning units, light bulbs and refrigerators.
- Raising import duty on inefficient appliances and reducing duty on energy-saving appliances, equipment and materials, including wind turbines and photovoltaic systems.
- Mandating PV systems and solar water heaters on all government buildings, and on all private buildings above a certain size including all hotels.
- Mandating the Ministry of Energy to collect data on greenhouse gas emissions from the power and transport sectors.

Building and Construction

- Revising building codes and enforcing construction norms that include energy efficiency measures such as thermal insulation and double glazing.
- Revising planning and permitting procedures so as to ensure that they comply with climate change concerns.

Transport

- Legislating stricter control of private passenger vehicle annual inspections.
- Mandating incentives and fiscal measures that facilitate the introduction of electric vehicles.
- Introducing fiscal measures incentivizing public transport, private buses and taxi services.

Water

- Establishing a high-level agency responsible for integrated water resources management, and defining its authority and mandate.
- Regulating the use of water from source to end-use point.
- Mandating wastewater treatment and prohibiting wastewater effluents into the ocean.
- Defining the norms for the reutilization of treated wastewater for irrigation.

Economy

- Introducing fiscal measures to promote energy efficiency by eliminating import duties and taxes on all energy-efficient equipment and materials.
- Set targets for vehicle efficiency.

National Adaptation Programs of Action

One of the requirements for less-developed countries (LDCs) signing up to the UNFCCC was that they were asked to prepare a National Adaptation Program of Action (NAPA). As mentioned in Chapter 1, only nine of the 51 SIDS are officially classified as LDCs, and were therefore required to prepare a NAPA. Each of them did so. Table 5.2 shows the sectoral focus of the projects proposed in the 12 NAPAs submitted. Note that Cabo Verde, the Maldives and Samoa have all 'graduated' from their LDC status since their NAPAs were submitted.

The priority sectors are clearly food security, coastal and marine ecosystems, and water resources. The health sector is in 5th position, and six of the 12 SIDS have no activities in this sector, which suggests that many SIDS are seriously underestimating the health impacts of climate change. The energy sector is evidently not seen as a priority, which may be due to the fact that the transition to renewable energy is viewed as more of a mitigation measure than a policy aimed at adaptation. None of the 12 SIDS sees insurance as a priority.

Table 5.2 Number of sectoral projects proposed in the NAPAs of LDC SIDS.

	Cross-sectoral	Food security	Coastal & marine ecosystems	EWS & disaster management	Education & cap. building	Energy	Health	Infrastructure	Insurance	Terrestrial ecosystems	Tourism	Water resources
Cabo Verde		2	1									1
Comoros		5		1			2	1		2		2
Guinea-Bissau		3	3	2	3					1		2
Haiti	1	1	3	1	1			1		7		2
Kiribati	2	1	3									2
Maldives	1	2	2				1	3				2
Samoa	2	1		1			1	1		1		1
Sao Tome & Principe	1	4		1	2	3	5	2		1	1	2
Solomon Islands	2		2	1				1			1	
Timor-Leste		2		1	1		1	2		1		1
Tuvalu		1	3	1				1				1
Vanuatu		1	1							1	1	1
Total sectoral projects	9	23	18	9	7	3	11	10	0	14	3	17

EWS: early warning systems.
Source: Adapted from UNFCCC (2016b).

Financing Adaptation

A considerable amount of money has been promised by the developed countries for actions aimed at fighting climate change: both by reducing emissions of CO_2 and other greenhouse gases, and by identifying and implementing the key adaptation measures necessary to reduce and cope with the worst impacts of climate change. Table 5.3 sets out the major multilateral climate funds and their basic features (CFU, 2016). There are also at least five bilateral funds as shown in Table 5.4.

Each fund has its advantages and disadvantages, and it will be one of the first tasks of a national climate change agency to develop an effective working relationship with the secretariats of these sources of funds. Several of the funds provide guidance and training on formulating proposals and applying for funding.

Apart from these sources of funds, UN agencies such as the UN Development Program (UNDP) and the UN Environment Program (UNEP) generally support protected area and marine ecosystem projects that fall within their specific mandates. It is essential that there is a person designated in the national climate change agency responsible for developing an excellent working relationship with these funding agencies including the bilaterals, the multilaterals, and the UN agencies.

In 2016, the Climate Fund Update website indicated that $146 million was approved for projects in small island developing states. Of this amount, 55% was for projects in

Table 5.3 Multilateral climate funds.

Fund	Administered by	Focus area	Date operational
Adaptation Fund (AF)	Adaptation Fund Board	Adaptation	2009
Adaptation for Smallholder Agriculture Program	International Fund for Agricultural Development (IFAD)	Adaptation	2012
Amazon Fund (Fundo Amazonia)	Brazilian Development Bank (BNDES)	Mitigation – REDD	2009
Biocarbon Fund	World Bank	Adaptation, Mitigation, REDD	2004
Clean Technology Fund (CTF)	World Bank	Mitigation – general	2008
Congo Basin Forest Fund	African Development Bank	Mitigation – REDD	2008
Forest Carbon Partnership Facility	World Bank	Mitigation – REDD	2008
Forest Investment Program	World Bank	Mitigation – REDD	2009
Global Environment Facility Trust Fund – Climate Change focal area (GEF 4)	GEF	Adaptation, mitigation, general	1991
Global Climate Change Alliance (GCCA)	European Commission	Adaptation, mitigation, general, REDD	2008
Global Energy Efficiency and Renewable Energy Fund (GEEREF)	European Commission	Mitigation – general	2008
Green Climate Fund	Green Climate Fund	Adaptation, mitigation, REDD	2015
Indonesia Climate Change Trust Fund	Indonesia National Development Planning Agency	Adaptation, mitigation, REDD	2010
Least Developed Countries Fund (LDCF)	GEF	Adaptation	2002
MDG Achievement Fund – Environment and Climate Change thematic window	UNDP	Adaptation, mitigation	2007
Pilot Program for Climate Resilience	World Bank	Adaptation	2008
Scaling Up Renewable Energy in Low Income Countries Program (SREP)	World Bank	Mitigation	2009
Special Climate Change Fund (SCCF)	Global Environment Facility (GEF)	Adaptation	2002
Strategic Climate Fund (SCF)	World Bank	Adaptation, mitigation, REDD	2008
Strategic Priority on Adaptation (SPA)	GEF	Adaptation	2004

REDD, Reduce Emissions from Deforestation and Forest Degradation.

Table 5.4 Bilateral climate funds.

Fund	Administered by	Focus area	Date operational
Australia's International Forest Carbon Initiative	Government of Australia	Mitigation – REDD	2007
Germany's International Climate Initiative	Government of Germany	Adaptation, mitigation, REDD	2008
Japan's Fast Start Finance (private and public)	Government of Japan	Adaptation, mitigation, REDD	2008
Norway's International Climate and Forest Initiative	Government of Norway	Mitigation, REDD	2008
UK's International Climate Fund	Government of the United Kingdom	Adaptation, mitigation, REDD	2011

the Caribbean SIDS, 32% in the Pacific SIDS, and 13% in the AIMS group of SIDS. Guyana, Samoa and the Maldives were the largest recipients, receiving between $70 and $80 million each. The most active fund was the Green Climate Fund (GCF). In 2016 this fund substantially increased its project approvals, which were evenly split between adaptation and mitigation (CFU, 2016).

The Green Climate Fund

The Green Climate Fund is a global platform designed to respond to climate change by investing in low-emission and climate-resilient development. The Fund was established with the aim of limiting or reducing greenhouse gas emissions in developing countries, and to help vulnerable societies to adapt to the impacts of climate change. Given the urgency of this challenge, the Fund is mandated to make a major contribution to the global response to climate change.

The Fund is accountable to the United Nations and guided by the principles and provisions of the UN Framework Convention on Climate Change (UNFCCC). The Fund is governed by a Board of 24 members, comprising an equal number of members from developing and developed countries.

The Green Climate Fund is the only stand-alone multilateral financing agency whose sole mandate is to serve the UNFCCC agreement, and to deliver equal amounts of funding to both mitigation and adaptation (GCF, 2016).

The Green Climate Fund is one of the largest funds. In 2016, the Green Climate Fund had approved a substantial amount of funding, with 19 new projects totalling US$1 billion. This value was equal to the combined project approvals of all other climate investment funds. In October 2016, the GCF board at its 14th meeting approved a further $745 million in funding proposals. The ten projects and programmes have a total value of US$2.6 billion and will help 27 countries to reduce their greenhouse gas emissions and adapt to the impacts of climate change. The board's target for 2016 is to approve US$2.5 billion in GCF funding (GCF, 2016).

Among the various projects approved for funding at the October meeting, only one was approved for a small island state: US$80 million was approved for the Sustainable Energy Facility for the Eastern Caribbean with the Inter-American Development Bank.

The GCF board also accredited eight additional organizations, bringing the total of Accredited Entities to 41. Among the eight organizations is the Caribbean Development Bank (CDB) based in Barbados.

The approved GCF projects cover renewable energy, water management, and several projects spanning more than one sector. The Sustainable Energy Financing Facilities project located in ten countries over the African, Asia-Pacific and Eastern European regions is the largest programme approved since 2003 and is funded through a grant of $34 million in addition to a concessional loan of $344 million. This programme aims to deliver climate finance to the private sector at scale though partner financial institutions.

Sub-Saharan Africa and Asia attracted the largest amounts of multilateral climate finance approvals: $630 million and $506 million respectively in the last 12 months out of a total of $2.8 billion. In Asia, 82% of new approvals have been directed towards mitigation projects. There has been a greater range of projects in sub-Saharan Africa, with 43% of new approvals going to adaptation projects, 39% towards adaptation projects, and 17% going to multiple focus projects (CFU, 2016).

Morocco is the largest recipient in the Middle East–North Africa (MENA) region, with five new projects totalling $44 million, most of which ($39.8 million) comes from the GCF for the Development of Argan Orchards in Degraded Environment project. Morocco has been dominating the funding allocated to the MENA region since 2003. The three largest projects in Morocco are funded by the Clean Technology Fund, including a $238 million concessional loan for the Nour II and III concentrated solar power investment programme.

In 2016, a total of $146 million was approved for projects for small island states. Of this amount, 55% was for projects in the Caribbean and 13% in the AIMS group of SIDS. Guyana, Samoa and the Maldives were the largest SIDS recipients, receiving between $70 and $80 million each.

An important attribute of the Green Climate Fund is that developing countries are able to gain access to funding with support from the Fund itself. The GCF provides early support for 'readiness and preparatory' activities to facilitate access to the Fund. The 'country readiness' funding is a dedicated and cross-cutting programme that aims to maximize the effectiveness of the Fund by ensuring access for countries that lack the institutional capacity to take fulladvantage of the availability of funds.

The readiness and preparatory support mechanisms are part of a continuing process to strengthen a country's engagement with the Fund. The GCF focuses its readiness support on countries that are particularly vulnerable – including the LDCs, the SIDS, and countries in Africa. A minimum of 50% of the funding set aside for readiness preparation is earmarked for supporting these countries (GCF, 2016).

The GCF has a total of $16 million available to provide support to countries and further funding may become available as required. Support is capped at $1 million per calendar year for a developing country. Support may be provided to countries directly through National Designated Authorities (NDAs) or Focal Points through a wide range of delivery partners with relevant expertise and experience. Alternatively, countries may select another institution that is familiar with readiness activities as its delivery partner. These institutions can be international, regional, national or subnational, public or private.

In November 2016, 57 developing countries were receiving readiness support from the GCF, with $16 million in readiness grants already approved. Of these countries, 37 were small island developing states, LDCs and African states.

In addition, the GCF has funding available to support prospective regional, national or subnational entities seeking accreditation with the Fund to prepare them to apply for accreditation, as well as to Accredited Entities to develop projects and programmes.

The Adaptation Fund

The Adaptation Fund became operational in 2010. The Fund has a number of features that distinguish it from other funding mechanisms: (i) direct access for developing countries to the resources of the Fund; (ii) an innovative source of funding; and (iii) its governance structure. The guidelines developed by the Adaptation Fund Board do not set out what kinds of adaptation measures are eligible, or which sectors have priority. However, the strategic priorities are defined as follows (Adaptation Fund, 2010):

- Supporting adaptation priorities determined by and within developing countries;
- Consistency with relevant national development, poverty reduction, and climate change strategies;
- Taking into account existing scientific and political guidance;
- Special attention to the particular needs of the most vulnerable communities.

The first step in accessing funds from the AF requires that a small island state (which, to be eligible, must have signed up to the Kyoto Protocol) submits an application for the approval of its Designated Authority (DA). The DA is an actual person and not a government agency. If a National Agency for Climate Change Action (NACCA) has been created at a supra-ministerial level (as recommended in this book), a senior member of the NACCA should be nominated as the DA. Their communications with the Adaptation Fund will be reviewed and approved by the NACCA, so effectively the NACCA becomes the contact agency for the Adaptation Fund, with the DA as its representative.

The role of the Designated Authority is specified by the Adaptation Fund as being:

- The government's focal point with respect to the Adaptation Fund Board and secretariat;
- Responsible for endorsing the accreditation application of a National Implementing Entity (NIE);
- Responsible for endorsing project and programme proposals submitted for funding.

Several SIDS have received funding from the Adaptation Fund since its inception in 2010. In a repor, the Adaptation Fund (2015) listed these grants as shown in Table 5.5.

What is interesting about these programmes is the role that the UNDP plays in the management of the funds. The UN agency acts as a multilateral implementing entity. Every small island state will have a UNDP office, and so an important option that should not be overlooked is to discuss with the UNDP country representative their possible role in acting as an implementing entity for the Adaptation Fund.

Table 5.5 Adaptation Fund support to the SIDS.

Project or programme title	Country	Grant, US$	Implementing entity	Approval date
Enhancing resilience of communities in Solomon to the adverse effects of climate change in agriculture and food security	Solomon Islands	5,533,500	UNDP	2011
Increasing climate resilience through an integrated water resources management programme in HA, Ihavandhoo, Adh, Mahibadhoo and GDh, Gadhdhoo Island	Maldives	8,989,225	UNDP	2011
Climate change adaptation programme in the coastal zone	Mauritius	9,119,240	UNDP	2011
Enhancing resilience of coastal communities of Samoa to climate change	Samoa	8,732,351	UNDP	2011
Strengthening the resilience of our islands and our communities to climate change	Cook Islands	5,381,600	UNDP	2011
Enhancing adaptive capacity of communities to climate change-related floods in the North Coast and Islands Region of Papua New Guinea	Papua New Guinea	6,530,373	UNDP	2012
Enhancing the resilience of the agricultural sector and coastal areas to protect livelihoods and improve food security	Jamaica	9,965,000	Planning Institute of Jamaica	2012

Source: Adaptation Fund (2015). Reproduced with permission.

The $5.38 million Cook Islands grant programme (Adaptation Fund, 2014) defined its goals as being:

- Strengthen capacity for all Cook Islands residents to address climate change challenges;
- Support adaptation and disaster risk reduction (DRR) at national level;
- Enhance resilience to climate change including weather- and climate-related disasters;
- Integrate climate risks and resilience into economic development.

Activities supported by the grant included:

- Establishing small-scale climate-resilient agriculture;
- Preparing integrated climate change adaptation and DRR action plans for 11 inhabited islands;
- Training stakeholders and key players in climate and disaster risk assessment and management;
- Piloting water infrastructure adaptation practices;
- Implementing climate-resilient coastal protection measures;
- Conducting and updating risk and vulnerability assessments at national level;
- Establishing a fisheries database to monitor changes in abundance.

Programme Development

The development of programmes that will attract funding for climate change adaptation is a complex process that can be conducted and completed in different ways. Several steps are necessary, some of which can be carried out in parallel. Table 5.6 outlines how a small island government might move forward towards an action plan for adaptation (not mitigation).

At this point, the NACCA has an outline for the key elements of government strategy concerning priority actions for adaptation, but there has been no discussion among the ministries themselves. The NACCA therefore convenes a meeting with all the ministry representatives to present the set of initial proposals. The aim is to validate the initial proposals and to identify synergies and cross-cutting actions where two ministries, for example, agriculture and environment, may need to cooperate and coordinate on the protection of a key watershed; or where the Ministry of Environment and the ministry responsible for fisheries will need to coordinate on the protection of a key fisheries area where coral reefs are being degraded and overfishing is a problem.

The aim should be to formulate a set of concrete priority actions that are now specified in terms of their location, expected results, and the ministries taking the lead on

Table 5.6 Outline of action plan for funding adaptation.

Initial steps	Tasks
1 Legally establish the NACCA	• Establish the legal mandate and authority of the NACCA to act as a policy-making and coordinating agency • Approve and mobilize the budget • Complete staffing and provide equipment, office space, communications and vehicles
2 Establish links to funding agency	• Identify the funds that *a priori* look appropriate for the main threats facing the country • Contact the secretariats of the different funds • Meet the requirements for establishing the NACCA as the National Designated Authority or Focal Point for the funds
3 Establish the Accredited Entity for the Green Climate Fund and/or the Adaptation Fund	• The Accredited Entity may be the UNDP, a financial institution like a development bank, or a well-established technical institution that can meet the fiduciary standards required by the fund
4 Meet with the ministries (individually) implicated in climate change action	• The ministries responsible for food security, agriculture, water resources, forests, coastal zone ecosystems, coastal infrastructure, fisheries, electricity generation, disaster response planning, and health, all have a role to play in defining climate change policy and the formulation of action plans for an adaptation response • Discuss the principal concerns for each ministry concerning climate change threats, and identify priority actions to respond to those threats • Each ministry should designate a climate change focal point to represent the ministry in future discussions

their implementation. At this point the government, represented by the NACCA, has a draft climate change adaptation plan, and all the relevant ministries are aware of it and informed as to their possible role and expected participation.

A consultative process now begins with non-governmental organizations that work in the sectors that will be negatively impacted by climate change: agriculture, food security, water resources (including watershed management), forests and terrestrial ecosystems, fisheries and the marine environment, coastal infrastructure and tourism, poverty alleviation, and health. A workshop format is proposed, led by the NACCA, where the government's draft climate change adaptation proposal is presented for discussion and review. It needs to be made very clear to all workshop participants that this is an initial draft subject to substantial revision if a consensus develops that this is required.

The result of this workshop is an agreement on a revised draft action plan that has been reviewed and discussed with the NGOs that work in the different sectors, and which may well have a role to play in the implementation of the final action plan.

The approach so far has been all top-down. The next phase involves consultation with the communities that will be involved in the implementation of the proposed actions. But this consultation should start with meetings with the regional administrative and local government structures. because these agencies must be fully on board and supportive of the action plan where it will be implemented in their regions and districts.

All countries have a government structure and agencies that operate at the regional and local level. There are departments and communes (as in Haiti), counties and parishes (as in Jamaica), or administrative regions with a governor or a préfet (as in Djibouti). Large towns will have their own administrative structures: a mayor and his or her aides, and managers responsible for schools, roads, water, sanitation, healthcare, electricity, and all the other services essential for the management and administration of the town and its urban environment. In addition, there will be community groups, community-based organizations (CBOs), non-governmental organizations (NGOs), and possibly private sector associations as well.

The NACCA consultative process starts at the top of the regional administrative structure: the governor, the préfet, or whoever administers the region where the proposed adaptation programmes may be implemented. After meeting with the head of the regional government structure, including the engineers and advisors who work at this level, the NACCA should convene meetings with local government agents to explain what is being proposed in their area and why.

These steps are followed by the full consultative meetings with local communities. These meetings need to be introduced and headed up by the local authorities – now fully aware of the projects that are being proposed and the logic behind them. The proposals are introduced by local government agencies who take the lead, with the NACCA engineers being invited to present the climate change context and the need for urgent action.

It is essential at this stage of finalization of the adaptation proposals that the concerns of the local communities are listened to and taken into account. Only the local communities know what is really happening to coastal ecosystems; they are the ones that witness the subtle shifts in the timing of seasonal events, sense the strengthening of slow-onset events, experience the changes in rainfall patterns and the strength of storms, and see the erosion at points along the coast.

Figure 5.3 illustrates the development process for a programme in Haiti aimed at developing an action plan employing an ecosystem-based adaptation approach to building resilience and strengthening coastal ecosystems. The approach employed was top-down *and* bottom-up.

- The top-down approach was aimed at ensuring the technical validity of the proposed actions. This requires accurate geo-referencing of ecosystem locations, followed by ground-truthing of ecosystem characteristics, the production of accurate maps that spatially configure the zone, and the development of technical recommendations by environmental and biodiversity experts.
- The bottom-up approach was aimed at ensuring that local government agencies and local communities are fully engaged and motivated. This requires that the risks from climate change are explained and understood, the need for action is agreed and supported at the community level, that communities and local government structures work together to identify problems and solutions, and have agreed on their responsibilities and roles as outlined in the proposals.

The top-down and bottom-up approaches meet in the middle – which is where the technical experts and the communities meet in a consultative process that works out the details of the proposed programme, ensures that the plan is feasible, acceptable and understood, and that the communities and local administrative structures regard the plan as 'theirs' (Bush, 2014).

At the same time that the consultative process is being carried out in the regional centres with local government and community representatives, the NACCA is occupied with discussing with the Accredited Entity and the participating ministries how the different projects are to be managed, and in particular who are to be the implementing or executing entities. The requirements differ according to the funding agency that is being solicited, but it is the responsibility of the NACCA to ensure that the projects are presented in a professional manner, that costs and benefits are accurately estimated, and that all the application requirements of the funding agencies are met.

This approach to programme development is only feasible if the NACCA is an independent agency with the authority to convene the different ministries. This is why it is not recommended that the NACCA be established within a ministry. Although this arrangement is quicker and less expensive to set up, it can make inter-ministerial cooperation and coordination difficult and ineffective.

It is also a mistake to believe that programme development should be driven entirely by an extensive consultative process, with no direction or guidance provided at the outset by ministry specialists or advisors who have good scientific knowledge of probable climate change impacts.

As an example of an extreme consultative approach, in 2015 the government of Palau published a comprehensive policy document called 'Palau Climate Change Policy: For climate and disaster resilient low emissions development' (SPC, 2015). This document, two years in the making, was the result of exhaustive consultative meetings held by ten working groups representing the following sectors:

1) Agriculture and fisheries
2) Health
3) Finance, commerce and economic development

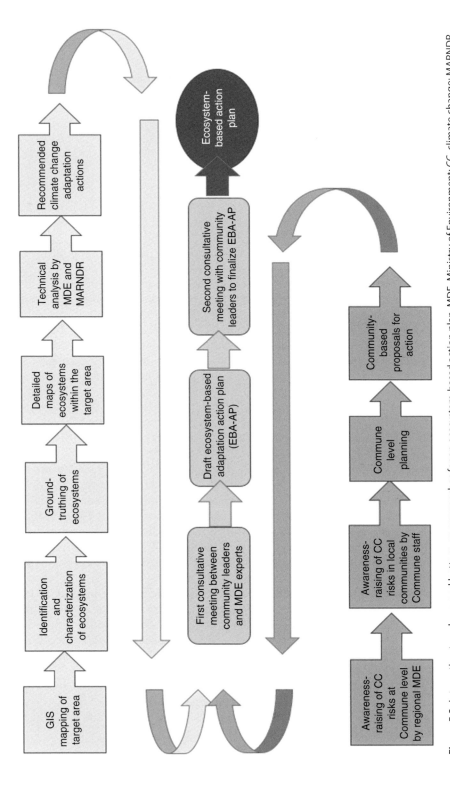

Figure 5.3 Integrating top-down and bottom-up approaches for an ecosystem-based action plan. MDE, Ministry of Environment; CC, climate change; MARNDR, Ministry of Agriculture. (*See insert for color representation of the figure.*)

4) Biodiversity, conservation and natural resources
5) Critical infrastructure
6) Utilities (electricity and water)
7) Society and culture
8) Good governance
9) Education
10) Tourism

The resulting 56-page full-colour report presents a long list of proposed actions organized into 16 sections. A total of 116 specific actions are listed in the plan. The budget for the whole programme is $500 million.

However, none of the actions are prioritized and few of the action items even indicate where the proposed activity is supposed to take place. Food security is not mentioned, although a plan for integrated water resource management is proposed.

The Government of Palau's climate change action plan is an extreme example of wishlist planning, where every ministry, local government agency, and community-based organization inserts everything they can think of into the programme, and all the concerns are dutifully added to the list and given equal weight. The result is a compendium of local concerns, none of which are sufficiently well developed to form the basis of an application to any of the international funding agencies. This result is much more likely to occur when climate change policy and action is entrusted to a committee, rather than an independent agency specifically charged with directing and coordinating climate change action, and provided with the legal authority and the financial resources to achieve its mandate.

Geographical Information Systems

An essential tool in the fight against climate change is GIS: geographical information systems. This capability allows agencies to use geospatial data to map and model climate change impacts on urban areas, watersheds, coastal zones, and near-shore marine areas using aerial photographs, satellite images and geodetic data. It is essential that all the small island states have access to these technologies, hire experienced and well-trained staff with the expertise to use them, and have the IT resources to fully exploit these new techniques. Small island states that lack these resources should immediately start discussions with donor agencies that can provide them.

In Haiti, the Centre Nationale d'Information Geo-Spatiale was set up by USAID several years ago, and has played a huge role in providing geospatial data to development programmes and helping them to plan their interventions and monitor longer term outcomes. It is impossible to plan effectively without maps and satellite imagery, and GIS products provide the platform on which climate change adaptation planning should be displayed and monitored.

Coastlines are a dynamic environment and GIS tools can be used to track geophysical changes, and to monitor projects aiming to strengthen the resilience of coastal ecosystems. In particular, the extent of forested areas including mangrove forests should be regularly monitored.

References

Adaptation Fund (2010) *Accessing Resources from the Adaptation Fund: The Handbook*. Available at www.adaptation-fund.org/documents-publications/publications/

Adaptation Fund (2014) Adaptation story: Cook Islands. Available at www.adaptation-fund.org

Adaptation Fund (2015) *Analysis of Climate Change Adaptation Reasoning in Project and Programme Proposals Approved by the Board*. AFB/PPRC.17/5. Available at www.adaptation-fund.org/wp-content/uploads/2015/12/AFB.PPRC_.17.5-Analysis-of-climate-change-adaptation-reasoning_final_Nov…pdf

Bush, M.J. and Sildor, E. (2014) Watershed Management in Haiti: Recommended Revisions to National Policy. USAID.

CFU (2016) Climate Funds Update. Available at www.climatefundsupdate.org/data/the-funds-v2. See also: blog.climatefundsupdate.org/2016/11/10/climate-funds-update-highlights-november-2016/

GCF (2016) Information on the Green Climate Fund, including the proposals accepted for funding from a number of SIDS. Available at www.greenclimate.fund

MARNDR (2011) *Politique de developpement agricole 2010–2015*. Ministere de l'Agriculture, Ressources Naturelles, et Developpement Rural, Port au Prince, Haiti.

MDE (2001) *Premiere Communication Nationale sur les Changements Climatiques*. Ministry of Environment, Haiti.

SPC (2015) *Palau Climate Change Policy: For climate and disaster resilient low emissions development*. Government of Palau, Secretariat of the Pacific Community.

UNFCCC (2016a) *Aggregate effect of the intended nationally determined contributions: an update*. Synthesis report by the secretariat. FCCC/CP/2016/2. Available at http://unfccc.int/resource/docs/2016/cop22/eng/02.pdf

UNFCCC (2016b) UNFCCC NAPA database. Available at https://unfccc.int/files/cooperation_support/least_developed_countries_portal/napa_project_database/application/pdf/napa_index_by_country.pdf

6

Country Profiles

This chapter reviews and discusses the adaptation programmes of each of the small island developing states, where these are outlined in their Intended Nationally Determined Contributions (INDC) communications to the UNFCCC secretariat, submitted for the most part in 2015. However, several SIDS did not submit an INDC communication because they are not a party to the UNFCCC. In this case, the review of their proposed adaptation policies and planning is based on sources easily available on the Internet: the AQUASTAT database managed by the FAO, the World Statistics Pocketbook series published by the UN Department of Economic and Social Affairs (UNDESA), the World Factbook published annually by the US Government's Central Intelligence Agency, as well as country-specific sources of information available on the Internet.

American Samoa

American Samoa is an unincorporated territory of the US located in the South Pacific Ocean, southeast of Samoa. Administered by the US Department of the Interior, the territory consists of five volcanic islands and two coral atolls, the largest and most populous of which is Tutuila. The Manu'a Islands (a cluster of three islands, Ta'u, Ofu and Olosega, located about 65 miles east of Tutuila), Swains Island (a small island with a population of less than 25 people) and Rose Atoll (an uninhabited atoll about 120 miles east of Tutuila) make up the rest of the territory. The combined area is 199 km².

The population of American Samoa has been declining over the past few years. In 2010 it was recorded as 64,000 (PCCP, 2015); in 2015 it was estimated as 54,000 (WF, 2015). The per capita income is a modest $8000, the lowest in the US. The island of Tutuila faces significant environmental and public health challenges that climate change will only intensify:

- almost 10% of residents do not have adequate indoor plumbing: piped water, a toilet, or both;
- leptospirosis, a serious waterborne disease, is widespread;
- heavy metals and other pollutants in the inner area of Pago Pago harbour make fish unsafe to eat (PCCP, 2015).

Climate Change Adaptation in Small Island Developing States, First Edition. Martin J. Bush.
© 2018 John Wiley & Sons Ltd. Published 2018 by John Wiley & Sons Ltd.

Typhoons are common from December to March. There are limited freshwater resources and, like so many small island states, the sustainable management of water resources is a priority issue.

Not being a party to the UNFCCC (because it is a territory of the US), American Samoa did not file an Intended Nationally Determined Contribution (INDC) communication in 2015. However, from the Pacific Climate Change Portal (PCCP, 2015), it is reported that the territory developed a Territorial Climate Change Framework (TCCF) with the aim of capturing projects, goals and strategies that increase its local capacity to develop and implement adaptation strategies to reduce vulnerability to impending adverse climate change impacts. Following this initiative, the territory established an associated Advisory Group to facilitate the implementation of the TCCF, mandated by an Executive Order of the Governor in 2011.

The Advisory Group is made up of six government agencies, a community college, two members of the government, and three mayors. The Advisory Group then established seven sub-committees to carry out thematic-based activities described as follows in the PCCP:

1) Coral reefs and mangroves – developing project plans related to enhancing coral reef and mangrove conservation and preservation with respect to climate change impacts.
2) Human health – developing project plans associated with improving human health issues that may be impacted by the effects of climate change. Examples of such issues include, but are not limited to, increased spread of disease and increased health problems resulting from changes in weather patterns.
3) Forestry, agriculture, and water resources – addressing issues related to enhancing the quality and sustainability of forest and water resources, as well as promoting sustainable agricultural practices.
4) Education and outreach – developing project plans related to improving climate change education and outreach opportunities throughout American Samoa communities. This may include the school curriculum, specific events, and other recommendations.
5) Coastal hazards – developing project plans to enhance American Samoa's resilience to climate change-related hazards. Examples of hazards include sea-level rise, increased coastal inundation, intensified storm systems, and others.
6) Development – developing project plans related to encouraging development throughout the territory. This will include addressing economic and business impacts expected from climate change.
7) Energy – addressing issues related to American Samoa's energy system, developing projects that encourage energy independence and sustainable energy use, and identifying options for alternative energy sources.

The management structure established in American Samoa with the objective to develop and implement actions aimed at adapting to climate change is a textbook case of how *not* to address climate change. Setting up a large multi-stakeholder group which then creates seven sub-committees is an exercise in futility. The result is what is called 'wishlist planning', where the main committee, after months of deliberation and dozens of sub-committee meetings, produces a long list of all the activities and projects that everyone agrees might possibly be useful. There is of course a budget, but one that just pencils in approximate cost estimates based again on wishful thinking. There is no prioritization, little realistic chance of funding, and almost no action.

The first step the territory should take is to establish a single agency with the authority to actually get things done – including producing and promulgating legislation that mandates action on priority concerns. These should include integrated water resource management and coastal and marine area zoning, protection and management. This agency should come directly under the Governor's office.

On the positive side, the territory has made significant progress in introducing renewable energy technologies. Several photovoltaic systems are up and running, including a 1.75 MW grid-connected array near the airport at Pago Pago. In 2008, the territory adopted a net metering law for distributed electricity generation up to 30 kW, allowing owners of small solar or wind systems to receive credit for excess power sent to the grid. In total, the territory has more than 700 kW of roof-mounted PV systems on government and private buildings, and a large solar water heating system at Tutuila's LBJ Tropical Medicine Centre (ETI, 2015).

Anguilla

Anguilla is a British overseas territory in the eastern Caribbean. One of the most northerly islands in the Lesser Antilles, the territory consists of the main island of Anguilla and several smaller islands and cays. The island's capital is The Valley. The total land area is about 90 km^2 and the territory has a population of close to 15,000.

In recent years, the island has become a popular tax haven, but tourism is now the mainstay of the economy. In 2014, over 70,000 tourists visited the island – almost five times the resident population. Agricultural production has declined to the point where its contribution to the economy is less than 3%. Almost all food products and fuel for electricity is imported.

Climate changes that are predicted for the island include:

- an increase in mean atmospheric temperature;
- reduced annual rainfall;
- increased sea surface temperatures;
- increased sea-levels;
- increased intensity of storms and hurricanes.

Like all the islands in the eastern Caribbean, Anguilla is vulnerable to extreme weather. The island has been hit hard by hurricanes: Luis (1995), Lenny (1999), Omar (2008) and Earl (2010) all caused considerable damage to coastal resources and infrastructure. Although the frequency of extreme weather may not increase in the Caribbean, the power of the storms is expected to intensify, due to higher seawater temperatures and greater levels of moisture in the atmosphere. All the Caribbean islands are hugely vulnerable to extreme weather, but islands that depend on beach tourism and which therefore have substantial investments in hotels and other tourism infrastructure on the coast are especially susceptible to damage from hurricanes. Hurricanes destroy not only built infrastructure; they can cause substantial erosion of beaches and damage to coral reefs – the natural resources on which tourism depends.

The analyses conducted by the Caribsave programme in 2012 assessed the most important impacts of the changing climate (Simpson *et al.*, 2012) as:

- coastal infrastructure and settlements due to sea-level rise and storm surges;
- impacts on community livelihoods, gender, poverty and development;
- agriculture and food security;
- energy and tourism;
- water quality and availability;
- human health;
- marine and terrestrial biodiversity and fisheries.

The Caribsave report (Simpson *et al.*, 2012) should serve as a model for other SIDS. The report identified the coastal areas most at risk from sea-level rise and storm surge by taking accurate GIS measurements along the coast and mapping the most vulnerable areas. This is the first step in planning for climate change on small islands. Although extreme weather cannot be avoided, its impacts can be mitigated by detailed and intelligent planning, and the first step is to know where the most vulnerable areas and the most exposed communities are located.

Anguilla has made progress in transitioning to renewable energy production and disaggregated electricity production. These measures also reduce the emissions of carbon dioxide, but their most important aspect is that disaggregated power systems are inherently more resilient than centralized systems. They also reduce the vulnerability of island economies to the volatility of fossil fuel prices when power generation is predominantly dependent on imported fuels – which was the case for Anguilla in 2015 when less than 1% of electricity came from renewable sources and there were no utility-scale solar or wind systems in operation (ETI, 2015). Clearly Anguilla is an island that could do better in this regard.

Antigua and Barbuda

Antigua and Barbuda are two islands in the leeward island chain in the eastern Caribbean. Antigua is the larger island: total area is about $442\,km^2$. The population is close to 100,000 persons. The economy is centred around tourism, although agriculture employs a significant fraction of the population. The climate is typical for the eastern Caribbean: temperatures range between 22 °C and 25 °C with little seasonal variation.

The agricultural production index was flat during the last three years, whereas the population has been increasing at about 1% annually. Agriculture accounts for only 2% of GDP, compared with 40% in the 1960s before the collapse of the sugar industry. Although small in GDP terms, the agricultural sector is still one of the largest employers: in 2013, 21% of economically active persons worked in the sector, or about 8000 persons. Sugar is still cultivated, and vegetable and fruit crops are grown on small farms, but Sea Island cotton is the only profitable export crop, according to the FAO database (www.fao.org/nr/water/aquastat/countries_regions/ATG/index.stm).

Tourism and its associated services now constitute the major foreign exchange earner: about 80% of GDP.

Like all the SIDS, emissions of CO_2 are very small – almost negligible on a global scale. Average precipitation is about 1050 mm annually with a dry season in the early part of the year; the wettest period is in September to November. The rural population is about 75% – relatively high for the Caribbean. Practically all electricity is produced from imported petroleum products, so tariffs are high: 0.37 $/kWh for residential customers and 0.39 $/kWh for commercial clients (ETI, 2015).

Islands in the eastern Caribbean have always had problems with finding enough water, and on most islands the situation is getting worse. Antigua and Barbuda have no significant surface water. Droughts occur every five or ten years, probably related to El Nino/El Nina cycles. Hurricanes (June to October) are a constant threat in the summer months; over the last ten years, four hurricanes have hit the islands. In an earlier period, three intense hurricanes passed directly over the islands: Donna in 1960, Luis in 1995, and Georges in 1998. The damage caused by Luis was estimated as equivalent to two-thirds of the country's GDP: a huge disaster for the economy.

Providing sufficient water for urban use, industry, tourism and agriculture in Antigua and Barbuda is a major problem. Desalination plants employing reverse osmosis and multi-stage flash distillation provide most of the country's fresh water, the largest of which is at the APAU facility at Crabbs. By law, all new homes are supposed to be equipped with rainwater collection and storage systems.

The changing climate brings both unpredictability and certainty: temperatures are certain to rise, as are sea-levels and ocean acidity. In the eastern Caribbean, it is expected that seasonal patterns of precipitation will become less predictable. and storms and hurricanes may become more intense and damaging. Pressure will intensify on all marine ecosystems due to increasing water temperatures and acidification. Water for urban areas, agriculture and coastal tourism will become more scarce and more difficult to manage. Low-lying coastal areas on the islands will become increasingly threatened by rising sea-levels and storm surges from extreme weather. Barbuda is particularly at risk: it is a low-lying coralline island. Its highest point is just 38 m above sea-level. A 1 m sea-level rise is estimated to impact 10% of major tourism resorts, all seaports, and 2% of major road networks.

The islands are among the top five countries most exposed to multiple hazards. All land areas and 100% of the population is exposed to two or more hazards. In terms of risk to GDP, Antigua and Barbuda is among the top 20 countries with an estimated 80% of GDP at risk from two or more hazards (GFDRR, 2012). After the devastation caused by hurricane Irma in September 2017, the island was declared 'almost uninhabitable'.

Antigua and Barbuda submitted an INDC communication in 2015 which detailed the country's adaptation actions as follows (INDC, 2015) – all planned for 2030 or before:

- Seawater desalination plants will increase their output from 5.4 million to over 8 million gallons/day.
- 100% of electricity demand in the water sector and other essential services (including health, food storage and emergency services) will be met through off-grid renewable sources.
- All waterways will be protected to reduce the risks of flooding and health impacts.
- All buildings will be improved and prepared for extreme climate events, including drought, flooding and hurricanes.
- An affordable insurance scheme will be available for farmers, fishers, and residential and business owners to cope with losses resulting from climate variability.
- The country will achieve an energy mix with 50 MW of electricity from renewable resources both on- and off-grid in the public and private sectors.

- All remaining wetlands and watershed areas with carbon sequestration potential will be protected as carbon sinks.

In the near term (before 2020), the country plans to establish efficiency standards for the importation of all vehicles and appliances.

While the programme of actions outlined in the INDC communication is a good start, other actions should be considered, including:

- Recycling urban wastewater for irrigated agriculture.
- Zoning and strictly managing all coastal areas so as to protect marine ecosystems, not just areas designated as 'carbon sinks'.
- Setting and applying energy efficiency standards for all new tourism infrastructure.

The country has good solar and wind resources that are not being fully exploited. Independent PV-hybrid minigrid systems are a cost-effective adaptation strategy and should be more strongly promoted. To encourage the development of independent power producers (IPPs), the utility has established interconnection standards for distributed renewable energy systems under 50 kW.

Planning for climate change action is vested in a Technical Advisory Committee (TAC) that is described in the INDC communication as an inter-agency, multi-stakeholder advisory committee that includes 15 government agencies, three NGOs and community interest groups, and one private sector coalition representative.

Establishing a technical advisory committee is not the best way to plan for and implement the urgent and transformational programmes needed to build resilience and successfully adapt to the threats posed by climate change. Responsibility for climate change policy and action needs to be assigned to an autonomous high-level agency directly under the President, with a clear mandate and the authority to prepare and promulgate legislation that regulates water resources and wastewater recycling, electricity production and IPPs, coastal and marine areas, and energy efficiency standards.

Aruba

The island of Aruba lies in the southwestern Caribbean. The country seceded from the Netherland Antilles in 1986 and became a separate autonomous member of the Kingdom of the Netherlands. The island is located about 1600 km west of the main part of the Lesser Antilles and 29 km north of the coast of Venezuela. Together with Bonaire and Curacao, Aruba forms a group referred to as the ABC islands. The island is flat and without rivers, but its many beautiful beaches and its warm and sunny Caribbean climate attract over a million tourists a year. The population is about 112,000.

Aruba does not feature in either the Aquastat database or the World Statistics Pocketbook of the UNDESA – nor did Aruba file an Intended Nationally Determined Contribution (INDC) communication to the UNFCCC in 2015.

The contribution of agriculture to GDP is miniscule. The economy is based almost entirely on industry (there is an important oil refinery) and tourism. The population density is high even before the tourists arrive. This almost certainly points to difficulties providing enough water.

Aruba has no surface water, so the island depends heavily on the desalination of sea-water. The first desalination plant was built in 1932, and more plants have been built as the demand for fresh water has increased in tandem with the burgeoning tourism industry. Desalination requires a lot of power, so the island's dependence on fossil fuels for the production of electricity is a serious burden on the economy. In response to these constraints, the government has promoted renewable energy technologies which now account for almost 10% of installed capacity.

Aruba lies out of harm's way when considering the threat of hurricanes and storms, and the relative unimportance of agriculture might lead one to think that Aruba has little exposure to the threats of climate change. But the country's dependence on coastal tourism is threatened by sea-level rise, coastal erosion, and the increasing stress on marine ecosystems including coral reefs. The country must therefore act swiftly and decisively to protect its coasts and marine ecosystems. Hotels and beach infrastructure (restaurants, bars, shops and clubs) should be strictly regulated so that all effluent and wastewater is treated according to international norms and standards that, if not already in place, must be introduced and enforced by the government.

With so many tourists on the beaches, coastal and near-shore marine areas will be under significant pressure. All coastal areas around the island should be zoned, and the regulations that apply clearly communicated to the private sector and enforced by the government. Hotels in particular require constant supervision. They should all be required to install rooftop PV systems and solar water heaters. These renewable energy systems should be considered an integral part of any hotel design. New construction should be set back from the high tide mark, and any remaining mangrove forests should be strictly protected including allowing space for them to retreat inland as sea-levels rise.

Aruba has made good progress towards shifting from fossil fuels to renewable energy. In 2012, the country announced a goal of 100% renewable energy use by 2020. In the same year the power distribution company, N.V. Elmar, instituted a revised net meter-ing policy that increased the maximum capacity for residential customers to 10 kW and compensates owners for surplus production. In addition, import duties on wind tur-bines, solar panels and electric cars have been reduced to encourage the use of these clean energy technologies (ETI, 2015).

The Bahamas

The Bahamas (or more correctly the Commonwealth of the Bahamas) consists of 700 islands and more than 200 cays, islets and rocks in the western Atlantic Ocean. Fewer than 30 of the islands are populated. The islands are spread over approximately $256,000 \, km^2$ of ocean with a total land area of $13,940 \, km^2$. Almost a quarter of the land area is less than 3 metres above mean sea-level. The population is about 390,000. The economy is mainly based on tourism; in second place are financial services. The bank-ing and finance sector accounts for roughly 15% of GDP. Agriculture and fisheries com-bined account for only 4% of GDP.

The population density appears low, but many of the smaller islands are uninhabited, so population densities on the main islands are much higher.

The Bahamas has designed and implemented a wide-ranging programme of adaptation measures intended to address the potentially disastrous effects of climate change. Water is once again the main focus of attention, as water shortages have been a serious problem in the Bahamas for decades. Only three islands have significant water sources. In 2006, more than 200 desalination plants were operating across the islands of the Bahamas. Club Med alone required $265 \, m^3$ of water a day (UNESCO, 2006).

A $45,000 \, m^3$/day reverse-osmosis (RO) plant has been supplying water to Nassau on New Providence since 2006. It is the largest diesel-powered seawater desalination plant in the world. Even more RO plants will be required in the future as rising sea-levels inevitably lead to saltwater intrusion into coastal aquifers.

The Bahamas lies in an area of the Caribbean where storms and hurricanes are expected to become stronger and more frequent. The protection of low-lying coastal areas where extensive tourism infrastructure is located is a massive but urgent challenge. Twenty per cent of the population is living on land just 5 metres above sea-level – around 75 000 people.

The Bahamas INDC communication, submitted in 2015, reviewed the adaptation measures undertaken up to that point in time and the different options available. For most of the sectors considered, the only action that had been taken was to *identify* the different options. Only for human settlements is there evidence of action. The INDC communication states:

> *Relocation of communities from the shoreline. This has already proved effective. New coastal defenses have been built and existing ones strengthened. Building codes have been made more robust to mitigate against increased wind loadings; and adapted to a loss of freshwater by employing reverse osmosis facilities throughout our islands to promote access to potable water.*

> (INDC, 2015)

These priorities make sense given the multimillion-dollar value of the coastal built infrastructure and the constant problem of providing enough water for residents and visitors.

In 2008, as part of the Caribbean Challenge Initiative, the Bahamas committed to the protection of 20% of the near-shore marine environment by 2020, and by 2015 was halfway towards achieving that goal (INDC, 2015). But much more should be done. *All* the coastal zones and inshore marine areas around the inhabited islands should be zoned and regulated.

Sea-level rise will eventually swamp many of the Bahamas low-lying islands, islets and cays. There is little that can be done to prevent this happening. A first step to managing this situation would be to assess the risk to each inhabited island and to identify the islands that are likely to be the first to become overwashed and uninhabitable. Not every island can be protected, and difficult choices concerning the allocation of resources will need to be made. Communities on the smaller inhabited islands will be forced to move and to settle on adjacent islands at higher elevations. Social tension and conflict are inevitable.

The Bahamas has made little progress in transitioning to renewable energy and reducing its dependence on heavy fuel oil and diesel. The Government shows little interest in moving away from a business-as-usual timeline. In 2008, the Government introduced incentives for solar technologies by reducing import duties on renewable energy items from 42% to 10% (ETI, 2015). Zero tax would have been a better move. Legislation defining the regulatory framework for net metering and net billing had not been finalized in 2015.

There is no indication in the INDC communication of how adaptation policy is being formulated or how the implementation of adaptation action is being implemented and managed. It has been mentioned before that adaptation programmes require strong leadership at the highest levels of government – *above* that of the line ministries.

Bahrain

The Kingdom of Bahrain is an archipelago of low-lying islands, islets, shoals and reefs situated off the southern coast of the Arabian Gulf. Saudi Arabia lies to the west and is connected to Bahrain by the King Fahd Causeway. The population in 2010 was about 1.2 million; two-thirds of that total are non-nationals. Bahrain has no natural surface freshwater resources. Rainfall is irregular and insufficient, and hydrocarbon resources are declining.

Bahrain has moved forcefully to implement measures to reduce CO_2 emissions by increasing the efficiency of energy use and by investing in renewable energy. Vulnerability and climate change impact assessments have been conducted, which led to the identification of four key sectors where adaptation measures are a priority: coastal zones, water resources, human health, and biodiversity.

Adaptation actions described in the INDC communication (INDC, 2015) focus on:

- Increasing coastal resilience to sea-level rise by dredging and reclamation.
- Implementing a climate-resilient integrated water resources strategy. A National Water Resources Council was established in 2008 to lead this programme.
- Strengthening food security by assisting the recovery of local fish stocks by landing artificial reefs in key zones.

Other actions being undertaken include water conservation, sustainable urban planning, and the rehabilitation of mangrove forests and seagrass beds. Sea-level rise is a major concern. About a third of the country's industry and built infrastructure is situated on the eastern coast of the main island – a zone vulnerable to flooding as sea levels rise.

But the major problem is once again fresh water. Almost 90% of Bahrain's water comes from desalination plants that are powered by natural gas. The amount of energy used to produce desalinated water in Bahrain has been progressively increasing over the past several years, rising by almost 50% during the period 2000–2010. In 2010, desalination accounted for as much as 30% of national energy consumption (Al-Zubari, 2014). Furthermore, the proven reserves of natural gas are limited. Estimates made in 2014 were that proven reserves were about 92 billion m^3, while annual consumption was close to 16 billion m^3/yr (WF, 2015). Clearly, Bahrain's natural gas is running out. Natural gas for desalination plants will soon have to be imported and the cost of producing water, already a major expense for the government, will become even more of a financial burden.

The adaptation programme is being led by a National Climate Committee. This organizational structure is unlikely to be strong enough to successfully manage the legislative actions required for Bahrain to tackle the problems of climate change.

A problem that is not mentioned in Bahrain's reports and documents related to climate change is the threat of extreme temperatures in the Gulf region that are expected to reach well over 45 °C. The country must urgently consider how best to cope with what will be increasingly common and life-threatening events. The energy required for air-conditioning systems and desalination plants is going to be substantial. This argues for introducing building codes that require extremely energy-efficient building design, and for the building of desalination plants powered by solar energy.

Bahrain has been slow to pilot and scale up renewable energy systems. A single 5 MW grid-connected system operates on the island. Offshore wind, which could be an option given the shallowness of the coastal marine zones, is not mentioned in the INDC report. The three fixed-axis turbines on the Bahrain World Trade Centre are an innovative feature but do not signal a serious attempt to develop the country's wind resources.

Barbados

Barbados is a Caribbean island country in the Lesser Antilles. The island is 34 km long and 23 km wide at its widest point, with an area of about 430 km^2. It is situated about 100 km east of the Windward Islands and 400 km northeast of Trinidad and Tobago. The island is a major Caribbean tourist destination. Being in the southern Caribbean, Barbados has generally avoided the worst of the Caribbean hurricanes, but it is still vulnerable to these violent storms. The population is close to 300,000, which gives Barbados one of the highest population densities in the Caribbean: 698 persons/km^2. This figure does not include the roughly half a million tourists who come to the island each year.

Barbados is a major tourism centre with substantial coastal infrastructure. Providing enough water for these visitors and the infrastructure that supports them is a major challenge. The island is ranked among the top 20 water-scarce countries in the world. The most severe impacts of climate change are the availability of water, agricultural productivity, increased land degradation, and reduced fish stocks. Reduced rainfall coupled with higher sea-levels will exacerbate saltwater intrusion into coastal aquifers. Other climate change impacts are likely to include periodic drought, flooding, storms, increased pest outbreaks, the spread of invasive species, and the increased probability of vector-borne and heat-related illnesses. Since the majority of the population is resident on the narrow coastal zone, the island is extremely vulnerable to climate change threats.

According to the country's INDC submission, the Government of Barbados is working to bring climate change adaptation into national plans and strategies. These include:

- Medium Term Growth and Development Strategy 2013–2020
- Physical Development Plan
- White Paper on the Development of Tourism in Barbados
- National Adaptation Strategy to Address Climate Change in the Tourism Sector
- Coastal Zone Management Plan

- Storm Water Management Plan
- Other sectoral plans for agriculture, fisheries, water and health.

National policy is set out in the National Climate Policy Framework which is monitored by the National Climate Change Committee (NCCC), a committee that includes government ministries, NGOs and private sector agencies (INDC, 2015)

Clearly a lot of planning has taken place, but one wonders what has actually been implemented and achieved. On this point the INDC is silent. This inherently sectoral approach, while it appears comprehensive, will inevitable suffer from a lack of overarching leadership, vision, and efficient allocation of scarce resources.

Like many SIDS, water supply and coastal zone protection are the key priorities. But these are cross-cutting priorities that do not fit easily into conventional line ministry programmes.

Barbados has built several desalination plants in an effort to ensure adequate supplies of potable water. The Spring Garden reverse-osmosis desalination plant was commissioned in 2000 and converts brackish water from ten wells. The desalinated water is pumped into an underground storage reservoir where it is mixed with chlorinated groundwater, bringing the water quality within international norms.

In recent years, Barbados has emerged as a global leader in the manufacture and installation of solar water heaters. Since 1970, over 50,000 units have been installed. Solar water heaters are a cost-effective technology that saves money for householders and reduces CO_2 emissions in regions where electricity is generated primarily from petroleum fuels, which is the case in Barbados.

It is interesting to note that the local industry really took off after the government introduced fiscal incentives in the national tax code to reduce the cost of the technology. Solar water heater equipment was tax-exempt; electric water heaters were taxed at a higher rate; and solar water heaters were mandated on all new public housing. This is a good example of how leadership and action at the highest levels of government can quickly mobilize the private sector and lead to rapid and important market penetration by renewable energy technologies (Ochs *et al.*, 2015).

Barbados has also forged ahead in introducing net billing procedures for residential PV systems. In 2014, the government reported that 424 residential PV systems were operating with a combined capacity of more than 3.5 MW (ETI, 2015). The island has made significant progress in mobilizing the transition to renewable energy technologies (BREA, 2015).

Belize

Belize is a country on the Caribbean coast of Central America, bordered on the north by Mexico and on the south and west by Guatemala. With an area of 22,800 km^2 and a population of about 370,000, Belize has the lowest population density in Central America. However, the country's population growth rate of 2.4% per year is one of the highest in the Western hemisphere.

Belize has plenty of water, so water supply is not a problem. In fact, more than half of the renewable energy electricity production is from hydropower. The problem is on the coast where over 20% of the population are living within 3 metres of mean sea-levels.

These are the areas where most of the tourism infrastructure is located. Managing the coastal zone and protecting it from rising sea-levels and storm surges should be the first priority for Belize. The country is often hit by hurricanes, and coastal flooding from storm surges and inland flooding from flash floods are likely to become more frequent with climate change. Belize is hit by a major storm on average every three years. Between 1935 and 2005, 11 hurricanes killed an average of 168 people per event (GFDRR, 2012).

Belize's INDC summarizes the components of its climate change action plan as follows:

- Enhanced food security and sustainability.
- Integrating climate change in a Revised National Forest Plan.
- Sustainable management of the fisheries sector.
- Adoption and implementation of the Belize Integrated Coastal Management Plan.
- Improved integrated water resource management.
- Integrate climate change in the tourism sector.
- Building resilience of human settlements.
- Enhanced resilience of the transportation sector.
- Strengthened and improved human health.
- Improved waste management.
- National Climate Resilience Investment plan.

This wishlist covers just about everything. Furthermore, according to the INDC submission, the document was submitted to the UNFCCC by the 'Climate Change Focal Point', with the proviso that a 'pending Cabinet consideration of the INDC … may be different to the one submitted' (INDC, 2015).

This arrangement strongly suggests that the government of Belize has done little more than delegate climate change policy to an advisor who was charged with drawing up a document that complies with Belize's obligations under the UNFCCC.

Apart from the urgency of protecting coastal areas, the other obvious vulnerability is Belize City itself. The city has been frequently affected by hurricanes. In 1931 Belize City was devastated by an unnamed Category 3 hurricane which resulted in 2000 deaths. The city was again hit in 1961 when Hurricane Hattie caused the greatest financial loss in Belize from a natural disaster to date. In response to these events, the government proposed to build a new capital at Belmopan, 50 miles inland, and to encourage the relocation of the capital. However, this plan was never completed (GFDRR, 2012).

The majority of the population across the country remains vulnerable because of: the relative lack of transport and flood protection infrastructure; high levels of poverty; concentration of urban centres in low-lying coastal areas; high levels of linguistic and cultural diversity; and poor access to information and healthcare. Belize's long low-elevation coastline holds approximately 45% of the population (including 20% in Belize City). Many of these coastal urban centres are actually below sea-level (GFDRR, 2012).

British Virgin Islands

The British Virgin Islands (BVI) is a British overseas territory located in the Caribbean to the east of Puerto Rico. The islands make up part of the Virgin Islands archipelago, which includes the US Virgin Islands and the Spanish Virgin Islands. The 150 km² territory consist of five main islands – Tortola, Virgin Gorda, Anegada, Peter Island and Jost

Van Dyke – along with many more smaller islands and cays. Only 15 or 16 of the islands are inhabited. The capital, Road Town, is situated on Tortola, the largest island. The country has a population of about 28,000, of which approximately 23,500 live on Tortola.

One of the most prosperous groups of islands in the Caribbean, the economy is highly dependent on tourism, which generates an estimated 45% of the national income. Around a million tourists, mainly from the US, visit the islands each year. Because of its traditionally close links with the US Virgin Islands, the British Virgin Islanders have used the US dollar as their currency since 1959. Livestock-raising is the most important agricultural activity.

Not being an independent country, the British Virgin Islands did not file an Intended Nationally Determined Contribution (INDC) communication to the UNFCC. However, the BVI government has issued a series of reports that have examined climate change impacts in detail, and set out comprehensive policies intended to mitigate the most dangerous and probable impacts. The goals and objectives of the country's climate change policy were outlined as follows:

1) *Natural resources and fisheries.* Enhance the resilience and natural adaptive capacity of natural resources, including terrestrial, coastal and marine ecosystems as well as the fisheries sector.
2) *Tourism.* Create and maintain a better-managed, more resilient, diverse and environmentally responsible tourism industry.
3) *Insurance and banking.* Minimize the vulnerability of insured and mortgaged properties to climate change impacts.
4) *Agriculture.* Strengthen food security by expanding local agricultural production and increasing its resilience to climate hazards/changes.
5) *Health.* Enhance the capacity of the healthcare sector and the public to deal with climate-related health impacts such as increased incidence of dengue, ciguatera and childhood asthma.
6) *Critical infrastructure.* Human settlements and water resources. Enhance the resilience of existing critical infrastructure and settlements to climate change impacts, while avoiding the construction of new ones in areas or with materials prone to climate hazards. Promote water conservation and efficiency while increasing resilience to flood events and drought.
7) *Energy.* Promote energy conservation and efficiency and encourage greater use of renewable energy to reduce the national energy bill and increase energy security.

What distinguishes this wish-list from many others is that it was followed up by very detailed proposals for accomplishing these goals (British Virgin Islands, 2011).

For all the Caribbean SIDS, the truth is that little can be done to reduce the threat of extremely violent weather, storm surges and coastal flooding. About 20% of the islands' land area is within 3 metres of mean sea-level, and thus seriously at risk of flooding from storm surges produced by hurricane-force winds. The protection of coastal infrastructure has to be an immediate priority, and not just for tourism. Communication systems and disaster response management systems are also critical, including the power supply systems that enable them function effectively.

Like the Bahamas, the British Virgin islands rely for potable water on desalination plants: seven are reverse-osmosis installations, and one is a multi-stage flash distillation unit (UNESCO, 2006).

Resilience in energy systems is increased by integrating renewable energy technologies into the energy mix, and by promoting independent power producers (IPPs) and residential PV systems. The territory appears to have done little in this regard. Although renewable energy installations (wind and solar) have been installed on the outer islands, little progress has been made on Tortola (ETI, 2015).

Cabo Verde

Cabo Verde consists of ten islands and eight islets in the Atlantic Ocean about 450 km west of Senegal. Six islands are in the northern group: Santo Antao, Sao Vincente, Sao Nicolau, Santa Luzia, Sal and Boa Vista; four are in the southern group: Maio, Santiago, Fogo and Brava. The population is about 525,000. The country is arid, with only 10% of land capable of supporting agriculture. Providing sufficient water for the indigenous population and the flourishing tourist industry is a major problem. Twenty desalination plants are operating on the main islands. Changes in the climate have already been noticed: more frequent extreme weather, sea-level rise, and degradation of fish stocks.

All but one of the islands are volcanic and rugged. Even so, almost of quarter of the land area is within 3 metres of mean sea-level and therefore seriously at risk from storm surge and flooding. The main climate threats are droughts and more intense storms leading to water scarcity, food insecurity, and disruptions in energy supplies and essential services. This argues for continuing investments in desalination technology, comprehensive management of water resources (including recycling wastewater), and a shift to more decentralized and independent energy systems so as to better cope with disruptions caused by intense and destructive weather.

Cabo Verde's INDC submission (INDC, 2015) identifies three key strategic areas for adaptation:

- Promoting integrated water resources management, guaranteeing stable and adequate water supply for consumption, agriculture, ecosystems and tourism.
- Increasing adaptive capacities of agro-silvo-pastoral production systems in order to ensure and improve national food production, and promoting Cabo Verde's ocean-based 'blue' economy.
- Protecting and preventing degradation of coastal zones and their habitat.

With so much expensive water being produced by desalination plants, the government should invest in sewerage systems and wastewater treatment plants that can collect and recycle wastewater to be used for agriculture and increased food security. In addition, according to the INDC report, fully a third of the population still cook with wood on three-stone fires. This segment of the population will be particularly vulnerable to the impact of drought and increased food insecurity.

Cabo Verde is committed to implementing the Sustainable Energy for All agenda, and the country hosts the ECOWAS Regional Centre for Renewable Energy and Energy Efficiency (ECREEE). The government's objective is 100% electricity production from renewables by 2030. Key measures include:

- Smart-grid enhancement for the country's nine independent power networks, with state-of-the-art power conditioning, production and distribution control.

- Increased levels of energy storage – batteries and flywheels.
- Constructing more renewable energy micro-grids.
- Facilitating more distributed residential PV systems.
- Systematic deployment of solar water heaters across all the islands.

Supporting measures include closer planning in public-private partnerships, simplified procedures for licensing and certification ('one-stop shops'), and the creation of more competitive market conditions and fiscal incentives to attract private sector investment (INDC, 2015).

Comoros

The Union of Comoros is an island nation in the Indian Ocean located at the northern end of the Mozambique Channel between northeastern Mozambique and northwestern Madagascar. The country consists of three main islands and numerous smaller islands. The main islands are Grande Comore (Ngazidja), Mohéli (Mwali) and Anjouan (Nzwani). A fourth island in the group, Mayotte (Maore), voted against independence from France in 1974, and is now an overseas department of France. The capital of the Union of Comoros, Moroni, is on Grande Comore.

With an area of $1660\,km^2$, excluding the contested island of Mayotte, the Comoros is the third-smallest African nation by area. The population, again excluding Mayotte, is estimated at about 800,000. The country is poor: about half the population lives below the poverty line.

The three main islands that constitute Comoros are constrained by inadequate transportation links and the lack of any mineral or fossil fuel natural resources. Agriculture, including fishing, hunting and forestry, accounts for 50% of GDP, employs 80% of the labour force, and provides most of the exports. Export income is heavily reliant on the three main crops of vanilla, cloves and ylang-ylang.

Water is scarce in many parts of the Comoros. Mohéli and Mayotte possess streams and other natural sources of water, but Grande Comore and Anjouan have few sources of water.

The impacts of climate change are much more severe for poorer countries where a large proportion of the population is rural and dependent on agriculture and fishing as a subsistence livelihood. Poverty and vulnerability go hand in hand, and programmes to reduce poverty will always strengthen the resilience of rural communities to withstand climate shocks and cope with extreme weather.

Comoros' INDC focuses on five key programmes, three of which are underway and two are planned (INDC, 2015):

- Reduce the vulnerability of agricultural systems to climate change and climate variability.
- Reduce the risks linked to climate change and the impact on water resources.
- Improve the integration of climate change threats in the coordination and monitoring of strategies, projects and planning.
- Reinforce the resilience of natural resources including watersheds and forests by diversifying livelihoods.
- Reinforce the resilience of vulnerable communities by strengthening their capacity to cope with climate change variability and natural disasters.

What is missing from this list is a focus on coastal zone management. Like most small islands, the majority of the population lives on the coast and is therefore exposed to storm surge and coastal flooding. Destructive cyclones have caused extensive damage to the islands in the past, and the changing climate is likely to increase the intensity of these extreme events.

Communications are an essential part of disaster preparedness, and the increasing use of mobile phones is an indication of how useful this technology is becoming for rural communities. The very low figure for per capita electricity consumption points to a high percentage of communities without electricity. Electricity powers communications and access to information. Small-scale photovoltaic kits and mobile phones are inexpensive, and are an important part of disaster preparedness; the government should encourage and facilitate their use.

Comoros desperately needs technical assistance and financial support. Funding can be accessed through several of the international funding sources that are available for climate change adaptation programmes: the Green Climate Fund and the Adaptation Fund, for example. But many poorer countries lack the technical capacity, the institutional experience, and the qualified manpower to access these sources of funding. What is urgently needed is technical assistance to help Comoros formulate the proposals and satisfy the eligibility criteria that will unlock the international funding for priority areas of adaptation and community disaster preparedness.

Cook Islands

The Cook Islands include 15 islands with a total land area of about $240\,km^2$. The northern islands are low-lying, sparsely populated coral atolls. The islands are governed in association with New Zealand, which is responsible for defence and foreign affairs. In 2011 the population was only about 18,000, the majority of which live on the main island of Rarotonga. However, the population of the country has been declining for several years mainly due to emigration to New Zealand and Australia. The country has become a popular tourist destination with many resorts and hotels. The principal town, Avarua, on the north coast of Rarotonga, is the capital.

Tourism is the mainstay of the economy, generating about 70% of GDP, mainly on Rarotonga. Agriculture, fishing, fruit processing, clothing and handicrafts are of lesser importance, although the financial services sector has become more prominent. About 75% of the outer island households engage in fishing, mostly for their own consumption. Agriculture provided 15% of GDP in 2000, but only 2.7% in 2010 (IRENA, 2013).

The country is hit regularly by cyclones. In 2005, an exceptional year, five cyclones swept across the islands, causing an immense amount of damage. The increasing investment in tourism infrastructure needs to be located and constructed so as to keep exposure to a minimum. Protecting coastal areas and strengthening disaster-preparedness systems should be a priority.

The Cook Islands submitted an INDC report (INDC, 2015) which proposes very little action aimed at strengthening the resilience of the island communities to cope with the negative impact of climate change. The outer islands are of low elevation and particularly at risk from sea-level rise and the increasing severity of cyclones.

The Adaptation Fund has provided over $5 million in grants to the islands. Activities have focused on:

- Establishing small-scale climate resilient agriculture.
- Preparing integrated climate adaptation and disaster risk reduction action plans for 11 inhabited islands.
- Training stakeholders and key players in climate and disaster risk assessment and management.
- Piloting water infrastructure adaptation projects.
- Initiating climate-resilient coastal protection measures.
- Conducting and updating risk and vulnerability assessments at a national level.
- Establishing a fisheries database to monitor changes in abundance.

This Adaptation Fund programme has correctly identified the key areas where climate change action should be a priority (Adaptation Fund, 2014). However, establishing a fisheries database to monitor changes in abundance, while certainly interesting, seems unlikely to provide actionable data (and fishers usually know where fish can be found and how catches are changing). It might have been better to establish mechanisms to monitor marine and coastal ecosystems including coral reefs, and to look at ways in which ecosystem resilience could be strengthened and maintained.

Islands need reliable electricity, and transitioning to renewable energy is not only cost-effective but helps build resilience into electrical power systems. Rarotonga, the main island, is still heavily dependent on diesel generators. Electricity is therefore expensive, averaging about 50¢/kWh for domestic users. In 2013, the country announced ambitious plans to have 100% of electricity generated from renewable energy (predominantly photovoltaic) by 2020.

Cuba

Cuba is the largest island in the Caribbean with a land area of $110,860 \, km^2$. More than 4000 islands and cays are located around its coasts. The southern coast includes the archipelagos of Jardines de la Reina and the Canarreos, while the northeastern coast is bordered by the Sabana-Camagüey archipelago which consists of over 2500 cays and islands. Cuba's population is only around 11 million, and the population density is about 100 persons/km^2, one of the lowest figures in the Caribbean among the larger islands. Long isolated because of the embargo imposed by the US, the relaxation of this policy in 2015 is encouraging increased tourism and investment. A fifth of the population is engaged in agriculture – a higher number than those working in industry.

All the Caribbean islands are threatened by hurricanes and severe storms, and Cuba is no exception. In addition, the country is estimated to have over 100,000 people living within 3 metres of mean sea-level, so the threat from storm surge and coastal flooding is serious. A substantial part of the population is dependent on agriculture and fishing – livelihoods which are likely to be negatively impacted by the changing climate.

Cuba's INDC submission (INDC, 2015) relates how changes in the island's climate have been observed and documented:

- Since the 1950s the average temperature has increased by 0.9 °C.
- The period 1990 to 2010 has been measurably warmer.

- Sea-levels have been rising over the last 40 years, averaging 1.4 cm a decade.
- Shorelines have been retreating on average by 1.2 metres a year.
- Rainfall in the dry season has been increasing, but the frequency and intensity of drought has also increased significantly since 1960, particularly in the eastern region.

Sea-level rise will have a major impact on Cuba. Many of the 4000 coastal islands and cays will be regularly flooded and eventually become uninhabitable. The Cuban government is well aware of the danger, and has been identifying and mapping vulnerable coastal zones for a decade (INDC, 2015).

Cuba's programme of adaptation action focuses on:

- Reducing coastal vulnerability for settlements threatened by sea-level rise and storm surge.
- Protecting mangroves and coral reefs.
- Integrating adaptation action into programmes aimed at food production, integrated water management, land management, fishing, tourism and health.
- Monitoring and measuring climate change to provide a basis for environmental decision-making.
- Reducing vulnerability in the health sector by developing better understanding of the links between climate change, climate variability and human health.

Cuba has well-organized and effective systems in place to protect communities from hurricanes and extreme weather, and it has been successful in protecting its people from vector-borne diseases like Zika. The country has also vigorously implemented energy efficiency programmes.

Dominica

Dominica is an island country approximately in the centre of the eastern chain of Caribbean islands. With an area of $750 \, km^2$, it is the largest of the islands in the Lesser Antilles island group. The population is small: only 72,000 people. The island is volcanic in origin and the topography is characterized by rugged and steep terrain running north–south along the centre of the island.

Like all the windward Caribbean islands, Dominica is constantly threatened by extreme weather during the hurricane season. Since the late 1970s the island has been increasingly affected by hurricanes and tropical storms. In August 2015, tropical storm Erika caused a huge amount of damage. The prime minister Roosevelt Skerrit said at the time:

> *Floods swamped villages, destroyed homes and wiped out roads. Some communities are no longer recognizable. The extent of the damage is monumental. We have in essence to rebuild the country.*

Erika caused almost US$400 million in damages – about three-quarters of the island's GDP.

According to Dominica's INDC, over the last 20 years economic development has been sluggish and there has been little improvement in the levels of poverty. With the decline in the cultivation of bananas (an important cash crop but one frequently devastated by hurricanes), many farmers began moving into the fishing sector, which now employs approximately 2000 registered fishers (INDC, 2015). More recently, the tourism sector has been developing, drawn by the island's ecotourism credentials.

Dominica has expended a huge amount of energy and time in developing strategies and proposing programmes to build resilience for coping with a changing climate. The planning process that produced the *Strategic Programme for Climate Resilience* (SPCR) and the 2015 INDC identified 11 priority adaptation measures that are to be implemented over the next five years. However, there is almost nothing in the INDC communication that discusses what the government has actually done in terms of adaptation measures aimed at increasing resilience. It is another example of wish-list planning and a donor-dependent mentality where government agencies produce reams of reports and documents but almost nothing actually gets done.

In 2015, the government moved to establish a Department of Climate Change, Environment and Development (DCCED). This new government department is intended to ensure the timely and effective implementation and coordination of the SPCR programme, and other programmes proposed in Dominica's *Low Carbon Climate Resilient Development Strategy* as well as serving as a National Implementing Entity in order to facilitate access to funding from the Green Climate Fund. The DCCED reports to a committee, the Council for Environment, Climate Change and Development (CECCD). This institutional structure guarantees that almost nothing will get done.

It is not clear that Dominica possesses the institutional capacity to implement and manage adaptation programmes. The *National Adaptive Capacity Assessment* conducted as part of the SPCR identified '**considerable limitations in climate change risk management capacity** at the systematic, institutional and individual levels, at the national, sectoral, district and local level, and within the public sector and civil society, highlighting the **need for considerable capacity building**' (INDC, 2015: emphasis is from the INDC document).

Dominica needs substantial amounts of international technical assistance to build institutional capacity and to help the government develop the expertise to access funding for adaptation programs that will build resilience. As if to underscore the urgent need for assistance, in September 2017, the island was once again hammered by a monster hurricane. Maria, a category 5 storm, absolutely devastated the island.

Dominican Republic

The Dominican Republic is a sovereign state occupying the eastern part of the island of Hispaniola in the central Caribbean. The western part of the island is Haiti. With an area of $48,400\,km^2$, the DR (as it is often called) is the second largest Caribbean nation after Cuba. The population is just over 10 million people, of which about a third live in the capital Santo Domingo. The Dominican Republic has the largest economy in the Caribbean and Central American region. The economy is dominated by services, particularly tourism. More tourists come to the DR than anywhere else in the Caribbean. GDP growth in 2014 and 2015 has averaged 7%, the highest in the Western hemisphere.

Like most of the islands in the Caribbean, the DR has been hit hard by hurricanes and tropical storms. With the changing climate the threat is likely to get worse. The main focus of adaptation programmes is therefore on the coastal and marine environment. While hotel managers fret about the eroding beaches, poorer communities that rely on fishing watch as sea-levels rise, coral reefs bleach out, and fewer fish are to be found.

Over 50,000 people live with 3 metres of mean sea-level. Santo Domingo itself is vulnerable to sea-level rise and flooding from storm surge. One of the poorest areas of the city, La Barquita, which sits close to the Ozama River, is frequently flooded. The community has now been moved to a new town, La Nueva Barquita, built by the government at a higher elevation.

The country's INDC submission is a modest publication – just three pages long. Adaptation is referred to as a 'constitutional priority' for the country. The sectors identified as most vulnerable are water, power generation, protected areas, human settlements and tourism. The elements of the strategic planning approach to adaptation include:

- ecosystem-based adaptation
- increased adaptive capacity and decreased territorial/sectoral vulnerability
- integrated water management
- health
- food security
- infrastructure
- floods and droughts
- coastal and marine areas
- risk management and early warning systems.

This list covers just about everything, but little is said about how this strategy is to be implemented and its objectives achieved; nor is anything said about how the adaptation programme is structured and managed.

In contrast to the implementation of adaptation measures, the pace of the transition to renewable energy in the Dominican Republic has been impressive. Several megawatt-scale PV plants are either operational or under construction. In total, over 200 MW of PV power is on the drawing boards. The DR has good wind, and over 85 MW of wind plants have been installed: Los Conas I and II (25.2 and 52 MW respectively) and Quilvio Cabrera (8.25 MW). Another 80 MW of wind power is at the planning stage. In addition, the country has at least 26 hydropower installations with a combined rated capacity of 580 MW (ETI, 2015).

The electricity sector has seen significant investment by the private sector since 1997, when the sector was reformed to allow private companies to participate in the generation and distribution of electricity. In 2012 there were 13 private companies generating power, and the largest generator is a private company, AES Andre, which produces about 15% of the country's electricity (ETI, 2015).

Fiji

The Republic of Fiji includes 332 islands, of which about a third are inhabited. The islands are volcanic in origin, and therefore rugged with some mountainous areas rising to over 1000 metres. Over 87% of the land is on the islands of Viti Levu and Vanua Levu.

Fiji has good soil and adequate rainfall and is well endowed with natural resources: timber, minerals and fish. Over 50% of the land is forested.

The population is estimated to be about 880,000. The economy, once resource-based, is moving towards tourism and services. But agriculture is still important: about 70% of the labour force works in the sector. Tourist arrivals rose from about half a million in 2005 to over 750,000 in 2015. The rugged terrain and good rainfall has been exploited for hydropower: about half the electricity production is from this renewable source of energy.

But the islands are frequently hit by dangerous storms. In early 2016, a cyclone named Winston swept across Fiji and killed over 40 people. It was called a 'monster' storm, packing winds of almost 300 km per hour. Winston was the most powerful storm ever recorded by authorities in the Southern hemisphere. The damage caused in Fiji amounted to about 10% of the country's GDP.

The Government of Fiji is aware of the danger and of the threat that climate change poses to the islands. Fiji's INDC outlines the progress that has been made towards adapting to climate change and building resilience. The government has conducted vulnerability and adaptation assessments for the whole of Fiji, invested in improving early warning systems, dredging of river mouths, construction of inland retention dams, and the construction of 'cyclone proof-homes' in the most vulnerable areas. The planting of mangroves, construction of seawalls and the relocation of communities to higher ground are part of what is referred to as 'ongoing adaptation initiatives' (INDC, 2015).

The first coastal community to be relocated to a higher elevation was Vunidogoloa in 2012. The government expects that other communities will soon need to be moved.

There is little that can be done to reduce the impact of powerful cyclones. At least satellite imagery allows governments to see them coming, to warn the population, move coastal communities out of harm's way, to set up emergency shelters, and to pre-position equipment and supplies necessary for disaster relief. In this context, autonomous local mini-grid power systems bring increased resilience and the ability to ensure that communication systems continue to operate in the aftermath of the kind of extreme weather that brings down power lines.

Fiji's INDC sets out the elements of an action plan aimed at meeting the key adaptation challenges:

- Establish a national platform for climate change and disaster risk management by 2015.
- Develop a National Strategic Plan for Climate Change and Disaster Resilience by 2015.
- Review the Fiji National Disaster Management Arrangements to include climate change by 2016.
- Review the National Building Codes by 2016.
- Development of local government self-assessment tool for climate change resilience by 2016.
- Review the town plan regulations to facilitate the enforcement of zoning and buffer zones for coastal areas, river banks, high-risk areas and mangrove areas, to be completed by 2016.
- Develop a comprehensive assessment framework, including adoption of the damage and loss assessment methodology by 2015.
- Develop hazard maps and models for all potential hazards (including sea-level rise, storm surge, flood and tsunami) by 2020.

- Integrate the climate change and disaster risk reduction into the National Development Plan by 2015.
- Revise capital budget appraisal guidelines to incorporate comprehensive hazard and risk management (CHARM) and vulnerability and adaptation assessment by 2015.

This list of actions is far too process-oriented, and most of these planning activities should have been completed at least five years ago. What is needed now is a clear programme of targeted action. There is funding available from several sources. Fiji should ask for international technical assistance and start to mobilize.

French Polynesia

French Polynesia covers over a hundred islands and atolls in the western Pacific to the east of the Cook Islands. There are five groups of islands: the Society Islands archipelago composed of the Windward Islands and the Leeward Islands, the Tuamotu Archipelago, the Gambier Islands, the Marquesas Islands, and the Austral Islands. About two-thirds of the islands are inhabited. Tahiti, in the Society Islands, is the most populous island and the location of the capital Pape'ete, a city of about 133,000 people. The total population of all the islands is around 282,000.

French Polynesia's tourism-dominated service sector accounted for 85% of total value-added for the economy in 2009, employing 80% of the workforce. A small manufacturing sector processes products from French Polynesia's primary sector, including agriculture, pearl farming, and fishing. About half of the territory is forested, but there are no protected areas.

As an overseas territory of France, French Polynesia did not file an Intended Nationally Determined Contribution (INDC), but like all the other small islands in the Pacific Ocean, the hazards are well known: more frequent and intense cyclones, rising sea-levels inundating low-lying atolls and eroding beaches, increasing periods of drought interspersed with increased rainfall, warmer temperatures reducing agricultural yields, and bleaching of coral reefs.

The most intense tropical cyclone to have impacted the area in the past few years was named Oli, which formed north of the Cook Islands in February 2010. Oli was one of the most devastating cyclones to have hit French Polynesia, forcing the evacuation of thousands of people and destroying over 280 houses in the Society and Austral islands. Between 1980 and 2012, eight cyclones triggered national natural disaster declarations in French Polynesia, causing the death of 33 people and major damage to housing, infrastructure and crops.

The methodology of adaptation planning is clear for small island states. Start with identifying the principal threats, then mapping the main areas of vulnerability and the level of risk involved, in terms of both people and monetary value. Then, for each high-risk and vulnerable geographical area, there needs to be a clear, detailed and specific programme of action aimed at building resilience and increasing the ability to cope with the threats presented by the changing climate. These action programmes are then reworked into specific projects that are eligible for funding from climate change funding sources.

One important way to build resilience and strengthen disaster-preparedness is to switch to decentralized power systems based on renewable energy. French Polynesia has significant hydropower resources, generating more than a fifth of electricity needs. The islands should invest in solar and wind power systems and reduce their dependence on imported petroleum fuels.

Grenada

The State of Grenada comprises three main islands: Grenada, Cariacou and Petite Martinique. The smaller islands are Ronde Island, Caille Island, Diamond Island, Large Island, Saline Island and Frigate Island. Most of the population lives on Grenada, the capital of which is St Georges. The largest settlement on the other islands is Hillsborough on Carriacou. The total population is about 106,000. The islands are of volcanic origin with rich soil. Grenada's topography is mountainous, with Mount St. Catherine being the highest point at 840 metres.

Grenada is a leading producer of spices: cinnamon, cloves, ginger, mace, allspice, orange/citrus peels, and especially nutmeg, are all important exports. Grenada is the world's second largest producer of nutmeg. But tourism is the main foreign exchange earner. Over 100,000 tourists arrive annually, many on cruise ships. Strong performances in construction and manufacturing, together with the development of tourism and higher education – especially at St Georges University – have contributed to economic growth, but the pace slowed in 2010–14, after a significant contraction in 2009 because of the global economic slowdown.

Like most of the Caribbean islands, Grenada is exposed to the violent storms of the Caribbean hurricane season. In 2004, Hurricane Ivan caused catastrophic damage to the island, severely damaging the agricultural sector – particularly nutmeg and cocoa cultivation – which had been a key driver of economic growth. The following year (2005) Hurricane Emily struck the island. But during the last ten years, Grenada has not been as severely impacted.

Grenada filed an INDC submission in 2015 which cites the impacts of climate change as increased incidence of droughts, longer dry seasons, shorter rainy seasons, increased temperature, coastal degradation, and intrusion of saltwater aquifers (INDC, 2015). The report discusses climate change adaptation in only very general terms. The government defines its priorities as:

- Enhancing institutional framework (which includes the National Climate Committee which according to the INDC has been 're-established').
- Building coastal resilience.
- Improving water resource management.
- Building the resilience of communities.

These are certainly the sectors on which adaptation programmes in Grenada should be focused. But the INDC report acknowledges the lack of institutional capacity that exists in the country. Grenada needs significant and long-term technical assistance in order for the government to formulate and present effective adaptation programmes eligible to be financed by the international funding agencies.

Guam

Located in the western Pacific, Guam is one of five American territories with an established civilian government. The capital is Hagatna, but Dededo is the largest city. In 2015, about 162,000 people were resident on the island. With an area of 544 km^2, Guam is the largest and southernmost island in the Mariana Islands archipelago.

The island is volcanic in origin with steep coastal cliffs and narrow coastal plains in the north, low hills in the centre, and mountains in south. The relatively flat coralline limestone plateau is the source of most fresh water. Almost half the island is forested.

US national defence spending is the mainstay of Guam's economy, followed by tourism and services. Total US funds directed towards the island's economy amounted to almost $2 billion in 2014, or 40% of GDP. Guam serves as a forward base for the US military and is home to thousands of American military personnel. Over a million visitors arrive annually (probably related to the substantial US military presence), and tourism accounts for 13.3% of GDP. Despite slow growth, Guam's economy has been stable over the last decade.

Guam has become the commercial hub of the Micronesian region, and is by far the most developed island in the area. Because it is a high island, Guam has good freshwater resources, and is not threatened by sea-level rise to the same extent as many low-lying island nations in the region.

Guam lies within the typhoon belt and is periodically struck by tropical storms and typhoons most frequently between June and December. In 2002, Supertyphoon Pongsona, packing winds peaking at 278 km/h, left the entire island without power and destroyed about 1300 houses. Damage on the island totalled over $700 million and was considered by the Federal Emergency Management Agency (FEMA) to be the costliest US disaster that year in a single state or territory.

According to the US Environmental Protection Agency (EPA), which has responsibility for Guam's environment, the island faces significant environmental challenges:

- Although water resources are good, infrastructure is in poor condition and there is a risk of contamination.
- The island's wastewater treatment plants have been chronically out of compliance with federal regulations and the country's own water quality standards.
- The additional population expected on Guam over the next several years will put additional stress on Guam's infrastructure and environment.

Climate change is likely to bring more extreme weather to the island. The focus of adaptation programmes has to be on risk reduction and developing increased resilience. The high population density, water stress and non-compliant wastewater treatment facilities all point to the need for a focus on integrated water resources management. The island's coral reefs in particular need much better protection.

Guam is totally dependent on petroleum fuels for electricity production. Renewable energy systems would provide better resilience against extreme weather. It is noteworthy that Typhoon Pongsona knocked out power over the entire island. Distributed power systems and mini-grids provide much better protection and ensure that communication systems and disaster management operations can function effectively. In 2013, the Guam Energy Action Plan was published. It says nothing about climate change, and makes only passing reference to renewable energy (GEAP, 2013).

Guinea-Bissau

The Republic of Guinea Bissau is a coastal nation in west Africa situated between Senegal and Guinea Conakry. The country covers 36,125 square kilometres, with a population of approximately 1.7 million. The topography consists of low-lying coastal plains with a deeply indented estuarine coastline rising to savanna in the east. Numerous offshore islands include the Archipelago Dos Bijagos, which has 18 main islands and many small islets. The country is very poor: two-thirds of the population lives below the absolute poverty line. Malaria is endemic and tuberculosis prevalent. Among the 51 SIDS, Guinea-Bissau has the lowest ranking in the UNDP's Human Development Index (ranked 183rd out of 188 countries listed) (HDR, 2015).

Guinea-Bissau is dependent on subsistence agriculture, the export of cashew nuts, and foreign assistance. The economy is based on farming and fishing, but illegal logging and trafficking in narcotics are reportedly important economic activities. The combination of limited economic prospects, weak institutions, and favourable geography have made the country a staging post for drugs bound for Europe (WF, 2015).

Guinea-Bissau filed an INDC report in 2015. The report notes that over the last 20 years the government has published 28 national plans and strategies, including three management plans, one action plan and one master plan in 2013 alone.

But little gets accomplished. The INDC report notes that human capacity is inadequate (and exacerbated by what it calls 'constant political and governance instability'). Institutions are weak; there is a lack of specialized staff in the field of climate change, and the State has weak financial capacity.

Guinea-Bissau is one of the most vulnerable countries among the group of small island developing states. It is poor, low-lying, heavily dependent on subsistence agriculture and artisanal fishing, and hugely exposed to all the negative impacts of climate change. A quarter of a million people live within 5 metres of mean sea-level.

Guinea-Bissau needs a substantial and urgent programme of international technical assistance to be initiated and supported by the UN agencies and their international partners. Money for climate change adaptation programmes is available from the international funds, but without the technical competence to produce the documentation and successfully complete and follow up on the application procedures, Guinea-Bissau will continue to be sidelined and marginalized.

Guyana

Guyana is a country on the northeast coast of South America, with an area of $215,000 \, \text{km}^2$. The population is approximately 800,000. Guyana has one of the highest levels of biodiversity in the world, with one of the most extensive mammalian fauna of any comparably sized country. About 80% of Guyana is covered by forest, with more than 1000 species of trees. This extensive forest biomass is a massive carbon sink and Guyana is a net absorber of carbon. Although rich in natural habitat and biodiversity, the country is otherwise poor; about a third of the Guyanese population lives below the poverty line and indigenous people are disproportionately affected.

The Guyanese economy is based largely on the export of six commodities – sugar, gold, bauxite, shrimp, timber and rice – which represent nearly 60% of the country's GDP.

These commodities are highly susceptible to adverse weather conditions and fluctuations in commodity prices.

The majority of the population lives on the coastal plains and is exposed to extreme weather, storm surges and flooding. Over a quarter of the population live with 3 metres of mean sea-level. However, the threat to Guyana's coastal population may not be high: no hurricanes have ever hit Guyana. But sea-level rise and more intense rainfall will gradually erode the coastal zone and push the population further inland. Periodic flooding caused by extreme rainfall is likely to be the principal threat to the population, although periods of drought and unpredictable seasonal rains will increasingly become the norm.

Guyana's INDC focuses mainly on its forests and how they are to be protected. Since Guyana is a net carbon sink, the preservation of its huge area of natural tropical forest is a global priority. The country also has ambitious plans for its transition to renewable energy: its target is 100% renewable energy by 2025. Given its substantial hydropower resources, this is not an unrealistic objective (INDC, 2015).

Guyana's adaptation programme appears to be secondary to the management of the country's forest resources. The programme commits to continuing 'basic work' on integrated water management infrastructure, the introduction of new agricultural techniques such as 'hydroponics and fertigation', and asserts that 'climate change considerations will be mainstreamed in all sectors of national development'. There is no vision, no objectives, and no plan.

A first step should be a detailed assessment of the coastal areas and the threats to the livelihoods of the coastal communities. Georgetown itself is only a few feet above sea-level. In fact, as noted earlier, the principal threat of floods is from torrential rain rather than storm surges and rising sea-levels. Georgetown was badly flooded in 2015, and these events can be expected to be more frequent and more intense.

Haiti

Sharing the smaller part of the island of Hispaniola with its more-developed neighbour, the Dominican Republic, the Republic of Haiti has struggled for decades with poor governance exacerbated by a succession of natural disasters that have often brought massive disruption to the fragile economy of the island. Although the country is relatively large for the Caribbean, with an area of $27,750\,km^2$, the population is now close to 11 million, giving a population density of 377 persons/km^2 – one of the highest levels in the Caribbean. A further 2.5 million Haitians are estimated to live abroad – an important source of remittances which amount to well over $1.8 billion annually, equivalent to about a third of the country's GNP.

Still the poorest country in the western hemisphere, about 60% of the population live below the poverty line, and about a quarter of the population are classified as extremely poor, living on less than $1 a day. Economic development has repeatedly been disrupted by external shocks including food and fuel price fluctuations and natural disasters. The event with the most devastating impact was the magnitude 7.0 earthquake in January 2010 which killed about 300,000 people and displaced 1.5 million in Haiti's capital and nearby towns, making it one of the deadliest natural disasters on record. The earthquake resulted in damages and losses of around $8 billion – 120% of GDP (SREP, 2015).

Poverty, corruption, vulnerability to natural disasters, and low levels of education for much of the population are considered to be Haiti's most serious impediments to economic growth. The government continues to rely on international economic assistance for fiscal sustainability, with over 20% of its annual budget coming from foreign aid or direct budget support.

Haiti's Ministry of Environment submitted its 'Contribution Prevue Determinée au niveau National' (CPDN, the French equivalent of an INDC) in September 2015 (INDC, 2015). In terms of adaptation, the CPDN defines the country's priorities as:

- integrated management of water resources and watersheds;
- integrated management of coastal zones and the rehabilitation of infrastructure;
- maintaining and reinforcing food security;
- information, education and awareness-raising.

While natural resources management, coastal zone management, food security and awareness-raising are all absolute priorities, these are sectors that have been priorities for decades. Almost nothing has been achieved. The CPDN confirms this by setting out objectives for the year 2030 that are essentially the same objectives that have been defined by the government for the last 20 years. The CPDN cannot be considered a serious document. After listing every possible and conceivable need – as proposed by the participants at all the planning meetings held with government agencies and community organizations at all levels – the document requests that $25 billion be made available for the implementation of the proposed programme of action!

This kind of wish-list planning is symptomatic of governments that pay only lip-service to their obligations under the UNFCCC convention. Planning for climate change adaptation and mitigation are routinely passed to the Ministry responsible for the environment, but this ministry in many countries – including Haiti – is one of the weakest, with only a modest budget wholly inadequate for implementing national level programmes aimed at reducing risk and building resilience. Climate change programmes are cross-cutting and require a substantial allocation of government funds. Larger and better-resourced ministries such as agriculture and forestry will resist changes to the status quo. Only with vigorous and sustained leadership from the President's or the Prime Minister's office will climate change programmes gain traction, realistic and well-designed programmes receive funding, and results in terms of increased resilience to climate change start to register.

First and foremost, Haiti needs a comprehensive technical assessment of the risks engendered by the changing climate. The risks need to be quantitative, detailed, and assessed at the level of each commune. What carries the highest risk: coastal zone flooding from storm surge or coastal zone inundation from watershed flash floods caused by violent storms? Which towns are most at risk from catastrophic flooding – Gonaives or Port-au-Prince? Which communities are most at risk? Those in the upper watersheds exposed to landslides and mudslides, or those in the lower watersheds exposed to flooding? What is the greatest risk to the low-lying quartiers of Port-au-Prince: disastrous flooding from watershed runoff, or violent storm surge intrusion from the sea? None of these questions appear to have been posed, much less answered.

Haiti needs long-term technical assistance in order to successfully design and implement a comprehensive programme aimed at adapting to climate change. The CPDN report acknowledges this need. It states:

> *The difficulty of accessing funding and appropriate technical resources, the lack of a regulatory and legal framework to facilitate climate change adaptation, consti-tute just some of the obstacles that prevent the implementation of the proposed activities. In this regard, the country needs institutional capacity building and technical and financial support in order to overcome these barriers.*

Substantial financial support from numerous international agencies is available for Haiti. But the country needs help to access these funds, and long-term capacity-build-ing reinforcement to assist the government to design and successfully implement prior-ity adaptation programmes.

Jamaica

Jamaica is an island country situated in the Caribbean Sea, and the third-largest island of the Greater Antilles. With an area of almost 11,000 km^2, it is the largest Anglophone island in the Caribbean with a population of about 2.8 million people and a population density of around 270 persons/km^2. Like the other Caribbean islands, Jamaica has been hit hard by extreme weather and hurricanes. Between 2001 and 2012 Jamaica was swept by 11 damaging storms, including 5 major hurricanes, and several serious floods. These events resulted in loss and damage estimated at J$128 billion. Hurricane Ivan in 2004 caused damage and loss evaluated at 8% of GDP. Hurricane Sandy in 2012 also caused extensive damage: the health, housing and education sectors suffered the worst damage (CCPF, 2015).

The government is well aware of the threat posed by the changing climate. The expected major impacts across Jamaica include:

- Temperatures increasing between 1.4 and 3.2 °C on average, with greater variability.
- Sea-levels rising between 0.28 and 0.98 metres.
- Less precipitation in the summer months (June–August).
- Greater frequency of flood events, landslides and droughts.
- Probable (>66% certainty) increase in hurricane intensity.
- More acidic and warmer seas; coral mortality (CCPF, 2015).

Jamaica filed an INDC submission in 2015 which talks more about planning for adap-tation rather than presenting what has actually been accomplished so far, or what is underway at the present time.

A more comprehensive document is the Climate Change Policy Framework, which was published in 2015. However, the Policy Framework is all about the process. The Ministry of Water, Land, Environment, and Climate Change (MWLECC) clearly has a wide-ranging and ambitious mandate. Since 2012, the government has assigned the responsibility for climate change to the MWLECC, established a Climate Change Division in the Ministry, appointed a Climate Change Advisory Committee, and estab-lished a Climate Change Focal Point Network 'to facilitate a multi-sectoral approach to climate change' (CCPF, 2015).

The climate change strategy has six key components:

1) Development of a climate financing strategy.
2) Development of research, technology, training, and knowledge management.

3) Regional and international engagement and participation.
4) Promotion of consultative processes and communication to improve public participation in mitigation and adaptation response measures.
5) Strengthening climate change governance arrangements.
6) Developing mechanisms and tools to mainstream climate change into ecosystem
   protection and land-use and physical planning.

The annex presents what it calls 'Key Policy Implementation Actions', but once again
every single proposed action (there are 27 of them) is about developing, enhancing,
organizing, planning, establishing and assessing things. Not a single concrete action is
proposed for a specific geographical location to address an identified threat posed by
climate change.

In contrast, energy policy in Jamaica is very much focused on well-defined priorities
and practical action. The government has been consistently moving forward with the
objective of transitioning to renewable energy, and backing up this policy with programmes to improve energy efficiency. Megawatt-scale wind power and photovoltaic
power systems, and decentralized PV systems have all been installed and successfully
operated. In 2015, the country had over 40 MW of wind power and 30 MW of hydropower installed, with more on the way (ETI, 2015).

Kiribati

The Republic of Kiribati is an island nation in the central Pacific Ocean. The nation
comprises 33 atolls and reef islands and one raised coral island, Banaba. The islands
have a total land area of 800 km^2, dispersed over 3.5 million km^2 of ocean. The permanent population is just over 100,000, about half of which live on Tarawa Atoll.

As an atoll country, Kiribati is almost entirely dependent on imported food and fuel.
Subsistence farming and fishing are the primary economic activities. Only about 18% of
the population has permanent employment, and half of them work for the government.
Tourism, so often the mainstay of tropical island economies, is almost nonexistent: only
6000 tourist arrivals were registered in 2013 (UNDESA, 2015).

Kiribati is one of the most vulnerable countries in the world to the effects of climate
change. Its low-elevation atolls, isolated location, and wide dispersion over large areas
of ocean make the country hugely exposed to rising sea-levels and storm surges.
Although most of the atolls are located at latitudes where storms are not frequent,
coastal erosion and inundation caused by spring tides, storm surges and strong winds
are regular events. Because the islands are small and many are long and narrow, the
entire population lives close to the ocean. In March 2015, many atolls were flooded and
seawalls and coastal infrastructure destroyed by cyclone Pam, the category 5 storm that
that devastated Vanuatu. As extreme weather intensifies in the region, Kiribati is likely
to experience these events more frequently.

In the longer term, the most serious concern is that sea-level rise will threaten the
very existence of Kiribati as a nation. But in the short- to medium-term, a number of
other expected impacts are of concern. These include water supply and food security.
As sea levels rise, seawater infiltrates into the subsurface freshwater aquifer, threatening
the water supply. At the same time, increased periods of drought disrupt and reduce
agricultural yields. The health impacts of deteriorating water quality can be severe,

particularly for children. Desalination plants have been built on Kiribati, but problems with maintenance and spare parts have frequently been a problem (SOPAC, 2011).

The 2014 Kiribati Joint Implementation Plan on Climate Change and Disaster Risk Management (KJIP, 2014) sets out a $75 million programme that identifies 12 strategic objectives, the most important of which (in terms of funding) is the promotion of sound and reliable infrastructure development and land management.

The harsh reality, though, is that nothing can be done to protect the Republic of Kiribati from rising sea-level and the increasingly strong impact of extreme weather. The atolls are too low-lying, too dispersed, and too poor. International assistance needs to focus on a coordinated exit strategy for a doomed country.

Maldives

The Republic of Maldives is a South Asian island country, located in the Indian Ocean southwest of India and Sri Lanka. The country consists of about 1200 tiny coral islands, of which only about 200 are inhabited. The country covers an area of approximately $90,000 \, \text{km}^2$ of ocean. Like Kiribati, the Maldives is one of the world's most geographically dispersed countries. The population is about 340,000. The country's economy is based essentially on tourism and fisheries. The number of tourist arrivals has been rising steadily over the last decade and is now over a million visitors annually.

The Maldives has been called the flattest country on Earth; the highest point on any of the atolls is 3 metres above mean sea-level (MSL), and most of the land is only 1 metre above MSL (USGS, 2016). Climate change threats to low-elevation coral islands and atolls are well known and documented. Sea-level rise coupled with increasingly extreme weather will drive many of the communities now living on the outer inhabited atolls over to the larger islands, including the already densely populated North Malé atoll. Although many of the islands are close to the Equator and therefore not often in the path of tropical cyclones, storms and heavy rain occur every year, and these events often result in serious flooding. The atolls are so low that ocean swells originating far from the islands can also cause widespread inundation. More than 90 of the inhabited islands experience annual floods (USGS, 2016).

The worse disaster was in December 2004 when the Maldives islands were devastated by the Indian Ocean tsunami; over 50 islands suffered serious damage, and 14 islands had to be evacuated. Several islands became uninhabitable, and over 20 resort islands were forced to close. The total impact was estimated at more than US$400 million.

The island of Malé is entirely protected by a massive seawall. Inhabited outer islands have varying degrees of coastal protection. Some that were hit hard by the 2004 tsunami are entirely protected, but to a lesser degree than the islands of Malé and Hulhule.

The Maldives filed an INDC in 2015 which presents a well-focused programme of action to strengthen the country's ability to cope with climate change. Proposed actions include:

- Enhancing food security, by establishing food storage facilities, and making local agriculture more resilient.
- Strengthening the resilience of infrastructure (including protecting the shoreline around the international airport, and moving the main commercial port from Male to Thilafushi).

- Measures to protect public health, mainly resulting from insufficient potable water and sanitation problems during flooding.
- Enhancing water security, by adding desalination capacity.
- Coastal protection of settlements and resort islands.
- Safeguarding coral reefs and their biodiversity through an ecosystem-based approach and by reducing coastal pollution.
- Protecting the tourism sector (although specific action is unclear).
- Protecting fisheries by diversifying and developing mariculture.
- Installing early-warning systems.

Cross-cutting measures include creating sustainable financing mechanisms, establishing a Maldives Climate Resilient Fund, and building institutional capacity to address climate change impacts.

These are all good proposals (apart perhaps for the Climate Resilient Fund – it might be more effective to develop the expertise to fully access the international funds available for climate change adaptation). What is clear, though, is that atolls that are frequently overwashed cannot sustain an island community: agriculture is impossible because of saline soils, trees die, and groundwater becomes brackish and useless. Uninhabited atolls will disappear; inhabited atolls will become uninhabitable, and atolls ring-fenced with hard defences will increasingly require frequent and costly maintenance, and regular construction programmes to increase their height and durability.

Marshall Islands

The Marshall Islands lie near the Equator in the Pacific Ocean, slightly west of the International Date Line. Geographically, the country is part of the larger island group of Micronesia. The country's population of about 55,000 is spread out over 29 coral atolls, including over 1000 individual islands and islets. About half the population live on Majuro where the capital is located. Twenty-two of the 29 atolls and four of the five small raised coral islands are inhabited. Almost all the population lives within a few hundred metres of the sea and less than three metres above it. Overwash and floods have destroyed homes and crops. Droughts of unusual intensity and length have necessitated emergency food and water drops.

In March 2014 almost 100 homes on the capital atoll Majuro were destroyed by a combination of high tides and 5-metre ocean swells running in from the northeast. More than 900 people were forced into shelters. Families returned to live in homes half collapsed into the sea. In the same year, US Geological Survey research showed that a mix of sea-level rise and ocean swell means 'many atoll islands will be flooded annually, salinizing the limited freshwater resources and thus likely forcing inhabitants to abandon their islands in decades, not centuries, as previously thought' (USGS, 2016).

Floods alternate with droughts. In early February 2016, the Marshall Islands declared a state of emergency to combat the extreme drought that threatened the region's freshwater resources for months. President Hilda Heine said the government was increasing initiatives to fight the effects of drought, including freshwater storage and installing reverse-osmosis desalination systems on several of the islands.

The INDC communication submitted by the Marshall Islands in 2015 speaks optimistically of the commitment of the country 'to the strongest possible efforts in safeguarding security and human rights, as well as advancing development aspirations, in light of climate impacts and risks'. A Joint National Action Plan sets out actions 'to adapt against the effects of natural disasters and climate change'. These include:

- Establish and support an enabling environment for improved coordination of disaster risk management and climate change adaptation.
- Public education and awareness of effective climate change adaptation and disaster risk management from the local to national level.
- Enhanced emergency preparedness and response at all levels.
- Improved energy security, working towards a low carbon emission future.
- Enhanced local livelihoods and community resilience for all Marshallese people.
- Integrated approach to development planning, including consideration of climate change and disaster risks.

What's missing here are concrete measures to safeguard the health of the islanders. Healthy people can better withstand environmental stress. Both floods and drought have widespread health impacts due to disruptions to drinking water supply and the lack of adequate sanitation.

The first desalination plant in the Marshall Islands was installed in the 1990s. During the 1998 El Nino, Majuro imported many small desalination units, most of which broke down because of lack of maintenance. In 2006 on Ebeye, two desalination plants were providing about $380\,m^3$ of water per day, supplying water to 38% of island households (SOPAC, 2011).

Most of the smaller islands in the Marshall Islands group cannot survive the changing climate. In the short term, desalination plants can at least provide clean drinking water. Hard defences can hold back the ocean for a while, but these are stop-gap measures.

Mauritius

The Republic of Mauritius is an island nation in the Indian Ocean about 1000 kilometres east of Madagascar. The country includes the two main islands of Mauritius and Rodrigues, and the outer islands of Agaléga, St. Brandon, Tromelin and the Chagos archipelago. The islands of Mauritius and Rodrigues form part of the Mascarene Islands, along with the island of Réunion. The population is about 1.3 million on a total land area of just over $2000\,km^2$. The capital and largest city is Port Louis.

Like all the SIDS, the country is vulnerable to the impacts of climate change: intense cyclones, abnormal tidal surges, prolonged droughts, flash floods, and increased sea surface temperatures are all expected to intensify. Natural resources are limited and there is increasing competition for land as the tourism sector continues to flourish. International visitors grew from 750,000 in 2005 to almost a million in 2013. The government's declared target is 2 million tourists annually.

As the demand for water from a burgeoning tourism sector continues to mount, the integrated management of water resources will become a priority. Compared with many Pacific Ocean SIDS, the country has good water resources, at least on the main island. On Rodrigues and the other islands, the situation is more problematic and desalination plants have been installed.

Coastal and marine ecosystems are exceptionally rich. Extensive reef systems surround all of the islands of the archipelago. Rodrigues in particular has an extensive reef which is three times the size of the island. The challenge for Mauritius is to protect and preserve the natural beauty of the islands while simultaneously building the resilience to cope and withstand the changing climate. A stricter regulation of the tourism sector is an absolute priority.

The INDC submitted by the government of Mauritius in 2015 sets out the priorities for adapting to climate change (INDC, 2015). The priorities are a wish-list of all and everything:

- Protection of infrastructure against 'environmental calamities'.
- Disaster risk management and resilience.
- Coastal zone management.
- Rainwater harvesting.
- Desalination systems especially for Rodrigues.
- Integrated pest and disease management.
- Efficient irrigation and climate-smart agriculture.
- Climate-smart fisheries.
- Improved marine and terrestrial biodiversity resilience.
- Mainstreaming adaptation in the health sector.
- Acquisition of hybrid and electric means of mass transportation.

The INDC report notes that the country has limited resources and is challenged by many pressing priorities such as free education, healthcare and the eradication of poverty. According to the government a 'Marshall Plan' on poverty alleviation was being prepared in 2015 (INDC, 2015).

But there is a lot that governments can do to protect coastal and marine ecosystems when there are a million tourist arrivals a year. Fiscal measures should ensure that tourists pay for their access to exceptionally beautiful but fragile ecosystems, and hotel operators and service providers need to be held in strict compliance with environmental laws and regulations.

Micronesia, Federated States

The Federated States of Micronesia is a multi-island country comprising four states – Yap, Chuuk, Pohnpei and Kosrae – spread across a wide area of the western tropical Pacific. The states include several hundred islands. The actual land area is only about 700 km^2, but the islands are dispersed over 2.6 million km^2 of ocean.

The topography varies from low-lying forested atoll islets, typically 1–5 metres above sea-level, to densely vegetated volcanic islands of several hundred metres elevation. On the main islands of each state there are modern developing communities, while on the atolls there are low-technology, traditional communities dependent on fishing, agroforestry, groundwater and rainfall. Throughout the islands the population lives in the coastal zone, and all the communities are vulnerable to climate-related changes in precipitation, sea level, storms and coastal erosion.

Climate and sea levels in the region are strongly influenced by the El Nino Southern Oscillation (ENSO). Under El Nino conditions the islands typically experience drought. Under La Nina conditions, the islands generally experience higher than normal sea-levels. Drought and marine inundation by high sea-level may damage soil, food resources and drinking water.

Drought can be a huge problem for the islands during El Nino years. In 2016, the lack of rainfall was especially severe. Small-scale reverse-osmosis systems were installed on many of the islands. In the El Nina years of 2007 and 2008, communities were flooded by a combination of large swells and spring high tides that eroded beaches, undercut and damaged roads, and intruded into aquifers and wetlands. Once again, food and drinking water were in short supply. A nationwide state of emergency was declared in December 2008 and food security was declared the top priority for the nation (Fletcher and Richmond, 2010).

How do we build resilience to climate change impacts in a country that consists of hundreds of islands dispersed over a wide area of ocean? Each island must first focus on managing water resources, and then ensure that fisheries are managed in the most efficient and productive manner. That means the strict protection of marine ecosystems and coral reefs. Overfishing can no longer be tolerated. Island communities must take the lead.

Cyclones cannot be prevented and are likely to get worse. Each island needs a disaster risk management plan: communications, community shelters, reverse-osmosis plants, and small-scale photovoltaic power systems all need to be cyclone-proof.

The Micronesia INDC submission is brief and uninformative (INDC, 2015), but the excellent report by Fletcher and Richmond (2010) provides sound advice and guidance:

1) *Education*: Managing climate risk can be facilitated with community involvement, but first the community has to possess the awareness and knowledge of climate risk.
2) *Community-based adaptation*: Each community should develop a shared vision of what is at risk, and what qualities to protect. This planning is the basis for regional planning and a national plan.
3) *Climate data monitoring*: Research is needed to improve regional databases that will improve models, forecasts and the assessment of risk.
4) *Geographical information systems (GIS)*: Marine inundation will worsen with sea-level rise. Remote sensing, aerial and satellite imagery, topographic and bathymetric data are needed to build digital elevation models to conduct vulnerability studies.
5) *Planning*: Create a national climate risk management plan that emphasizes community-based adaptation.
6) *Food and water resilience*: Collect data on food and water resources and trends. With international assistance, build knowledge of tropical agro-forestry practices. Monitor sea-level and rainfall to forecast events when food and water assistance will be needed (Fletcher and Richmond, 2010).

Montserrat

Montserrat is an island in the Caribbean in the chain of the Lesser Antilles with the status of a British Overseas Territory. The island is small, with a tiny population of only 5000. Most of the population left the island in 1995 when the Soufriere Hills volcano violently erupted, destroying the capital city, Plymouth. Between 1995 and 2000, two-thirds of the island's population left, primarily going to the UK. The volcano is still active and a large part of the island is off-limits to visitors.

Like all the other islands in the Lesser Antilles, Montserrat is vulnerable to intense storms and hurricanes. Changing patterns of rainfall will disrupt traditional agricultural practice, and water will need to be managed much more carefully. Although the Soufriere Hills volcano remains a quiet but looming threat, tourism is picking up. Coastal development must be strictly regulated if beach erosion is to be minimized and marine ecosystems safeguarded. With such a small population, it should be possible to mobilize community action relatively quickly.

Montserrat has organized workshops and held a series of consultative meetings, but little action seems to have been taken. As an Overseas Territory, Britain has primary responsibility for increasing Montserrat's resilience to climate change threats. Because of its status, the country did not prepare an INDC communication to the UNFCCC secretariat. Information is therefore limited on what action the British government is proposing (Wade *et al.*, 2015).

Small islands often state that they lack the financial resources and experience to manage the technical complexities of climate change. But building codes, environmental regulations and coastal zone management plans do not require substantial funds to develop and promulgate. The enforcement of the legislation requires forcefulness, determination and community support. Small island government can do a great deal and make substantial progress while waiting for financial support from climate change funding sources.

Nauru

The Republic of Nauru is one of the smallest independent states, home to only 10,000 persons on an island of $21\,km^2$. It is a coral island, isolated in the Pacific Ocean 42 km south of the Equator and about 1300 km west of the International Date Line. The island has been largely ripped apart by the intensive strip-mining of phosphate – an industry which for a while 50 years ago made Nauru one of the richest countries on the planet.

The mining of this mineral destroyed large swathes of original forest and arable land, and this devastated area was never returned to its natural state. The scarcity of arable land and freshwater resources, coupled with the island's isolation, its dependence on imports for food and energy, its degraded environment, and the emergence of chronic health problems, make achieving sustainable development for this island state an almost impossible task (INDC, 2015).

More recently, Nauru has been in the headlines because it is where Australia locates its 'regional processing centre' for asylum-seekers. This is a convenient location for the Australian government – far from its shores. In return Australia provides foreign assistance upon which Nauru is almost entirely dependent.

The Government filed an INDC submission in 2015 that correctly judged that action to reduce the island's miniscule emissions of greenhouse gases were effectively a waste of time and money. Instead, the government has focused on the need to accelerate the transition from imported petroleum fuels for electricity generation to renewable energy power systems. This transition results in a reduction in CO_2 emissions, so mitigation is an important co-benefit arising from this policy.

However, Nauru faces all the hazards presented by climate change. The island is experiencing coastal erosion and declining productivity of its extensive coral reefs, rising ocean temperatures, ocean acidification, sea-level rise, and an increase in the number of intense storms and droughts will further damage the island's ecosystems. Climate-related disasters have already had huge impacts on economic growth and national development (INDC, 2015).

The focus for Nauru has to be the protection of its coastal and marine environment and the careful management of its water resources. As an independent nation, Nauru has access to international funds that are specifically for climate-related action focused on adaptation. The island needs immediate help to access these funds, and Australia would seem the ideal partner in this regard.

New Caledonia

New Caledonia is a French overseas territory (called in French a 'collectivité unique') in the southwest Pacific Ocean. Lying to the southwest of Vanuatu in the Southern hemisphere, the extensive archipelago includes the main island of Grande Terre, the Loyalty Islands, and the small uninhabited dependencies of Walpole Island, the d'Entrecasteaux reefs and the more distant Chesterfield Islands.

Grande Terre is the sixth largest island in the southwest Pacific. The territory has a total land area of $18,575\,km^2$ and a population of close to 270,000. The capital of the territory is Nouméa.

The mineral resources of New Caledonia are immense. Nickel and chrome are the most important minerals, and nickel is the leading export. Large deposits of iron, manganese and cobalt have been discovered, and deposits of antimony, mercury, copper, silver, lead and gold are also present. So the mining sector – with all its associated environmental impacts – is expected to continue to develop. But tourism is also flourishing, with around 100,000 tourists now arriving each year. The territory has a large network of marine protected areas that includes extensive areas of coral reef ecosystems: New Caledonia has a larger area of coral reefs than any other South Pacific territory.

New Caledonia is vulnerable to extreme storms with generally 2 or 3 cyclones passing close to the territory each year. In 2015 Cyclone Pam, one of the strongest cyclones ever recorded, inflicted massive destruction on Vanuatu before passing just to the east of New Caledonia. These destructive storms are expected to intensify with the changing climate.

The Secretariat for the Pacific Regional Environmental Programme (SPREP) has set out the Climate Change Strategic Priority as:

> *By 2015, all Members will have strengthened capacity to respond to climate change through policy improvement, implementation of practical adaptation measures, enhancing ecosystem resilience to the impacts of climate change and implementing initiatives aimed at achieving low-carbon development.*
>
> (SPREP, 2017)

But the Pacific SIDS should be much further along than this. By the end of 2016, each of the island states should have approved and implemented a well-defined and focused climate change action plan, and should have established a working relationship with the Green Climate Fund and the other sources of funding for climate change adaptation.

The major challenge for New Caledonia is to manage effectively its marine protected areas with their extensive coral reef ecosystems, to ensure their ability to cope with the stress of higher sea surface temperatures and seawater acidification. If the mining sector dumps tailings and other liquid effluents into the coastal environment, this stress, coupled with the impacts of climate change, will devastate fragile marine ecosystems.

New Caledonia has hydropower, megawatt-scale wind power, and photovoltaic systems on the smaller islands; for instance, electricity on Tiga island is 100% photovoltaic. The development of the mining sector will rapidly ramp up the demand for electrical power, and the challenge for the country is how to manage this growth sustainably with minimal environmental impacts on protected areas and with reduced levels of CO_2 emissions.

Niue

Niue is a small island in the Pacific Ocean located partway between Tonga, Samoa and the Cook Islands. Its area is about $260 \, km^2$, making Niue the world's largest elevated coral atoll – its highest point is 70 m above sea-level. The country includes two reef atolls: Anitope and Beverridge, where commercial fishing is banned. The capital is Alofi on the western side of the island. The population of only about 1600 people live in 14 villages dotted around the island close to the coast.

The island is vulnerable to climate risks such as tropical cyclones and droughts, geological risks such as earthquakes and tsunamis, and human-caused risks such as disease outbreaks and contamination of its freshwater supply.

The island has no surface water and relies upon groundwater resources and rain catchments. Groundwater is recharged via rainfall infiltration, but the porous soil renders the underground fresh water vulnerable to contamination from both human and natural causes.

The economy is heavily dependent on New Zealand, which has a statutory obligation to provide economic and administrative assistance to the country. Aid accounts for 70% of GDP, and 56% of those formally employed. Secondary sectors include tourism (7000 visitors in 2013) and fishing. Noni and vanilla are being developed as export crops (JNAP, 2012).

Niue filed an INDC report in 2015 that set out a likely scenario for the future to 2100:

- El Nino and La Nina events will continue to occur, but there is little consensus on whether these events will change in intensity or frequency.
- Annual mean temperatures and extremely high daily temperatures will continue to rise.
- Mean annual rainfall could increase or decrease (the models are unclear), but more extreme rain events are likely.
- Ocean acidification is expected to continue, and the risk of coral bleaching will increase.
- Sea-levels will continue to rise.

Climate change impacts are likely to further intensify freshwater lens and coastal water quality issues for the island. For these reasons, protecting and enhancing natural resources, adequate sanitation, and wastewater treatment are among the government's main priorities (INDC, 2015).

The government has recognized that transitioning to renewable energy has multiple economic benefits for small islands, and that this policy will also reduce CO_2 emissions.

However, the island is a net sink of CO_2, so it is debatable whether scarce resources should be focused on trying to reduce CO_2 emissions in the transport sector – better to focus on coastal zone protection and integrated water management.

Northern Mariana Islands

The Northern Mariana Islands, together with Guam, make up the Mariana Archipelago in the Pacific. The US-administered territory consists of 14 islands in a chain running roughly north–south over a distance of about 600 km. The southern islands are lime-stone with fringing coral reefs, while the northern islands are volcanic. The northern islands are all either uninhabited or only very lightly populated, in part because there are active volcanoes on three of the islands: Anathan, Paga and Agrihan.

The southern islands, Saipan, Tinian and Rota, are the largest and most densely popu-lated islands in the chain, accounting for 65% of the aggregate land area, 99% of the population and almost all of the economic activity and energy consumption.

The Pacific islands are highly influenced by El Nino and La Nina events and the Northern Mariana Islands are no exception. These ENSO cycles create pronounced variability in storms and precipitation.

The islands are located in the so-called Typhoon Alley, directly north of Guam, and one or more of the islands are hit by typhoons each year. In 2002, Rota was pummeled by Supertyphoon Pongsona which packed winds of 130 kph and pushed up a storm surge of 6.7 metres (RISA, 2016). Then, in 2004, Typhoon Chaba inun-dated many of the southern Mariana islands, dropping 20 inches of rain near the airport on Rota.

Coastal protection and integrated water management are once again the order of the day. Although tourism has dropped markedly over the last ten years, it is likely to pick up once again given the natural beauty of the islands. Strict regulation of coastal construction and full compliance with building codes and environmental regulations are essential. A rapid shift to renewable energy should be a priority, given the islands' reputation as the sunniest in Micronesia.

Palau

Palau is a multi-island country located in the western Pacific Ocean. The country includes approximately 250 islands that together form the western chain of the Caroline Islands in Micronesia. The capital Ngerulmud is located on the island of Babeldaob. The islands are sparsely populated; the total population is about 21,000, most of whom live on the island of Koror.

Like many Pacific Ocean islands subject to frequent extreme weather, the worst climate change impacts are the degradation of the coastal and marine environment, and the increasing frequency of droughts. In early 2016, Palau declared a state of emergency because of worsening drought conditions. Drought and unpredictable precipitation coupled with increasing temperatures invariably threaten food security, so these sectors are generally the priority for Pacific island SIDS.

Palau filed an INDC communication in 2015 but it is very brief and not very informative about adaptation measures. However, in 2015 the Government of Palau published a comprehensive policy document called *Palau Climate Change Policy: For Climate and Disaster Resilient Low Emissions Development* (SPC, 2015). This document, two years in the making, was the result of extensive consultative meetings held by ten working groups representing the following sectors:

1) agriculture and fisheries
2) health
3) finance, commerce and economic development
4) biodiversity, conservation and natural resources
5) critical infrastructure
6) utilities (electricity and water)
7) society and culture
8) good governance
9) education
10) tourism.

The 56-page full-colour report presents a long list of proposed actions organized into 16 sections. A total of 116 specific actions are listed in the plan. The budget for the whole programme is $500 million. However, nothing is prioritized, and few of the action items even indicate where the proposed activity is supposed to take place. Food security is not mentioned, but at least a plan for integrated water resource management is proposed. *Not a single action item is explained or justified in any detail.* This is an extreme example of wish-list planning.

Papua New Guinea

Papua New Guinea takes up the eastern half of the island of New Guinea and its offshore islands in Melanesia, a region of the southwestern Pacific Ocean. It is the largest country in the Pacific region in terms of area.

One of the most culturally diverse countries in the world, 80% of the population of over 7 million people live in traditional villages and have few modern facilities. Most communities depend on the extensive tracts of primary forest for hunting, plant foods, and traditional medicine. Climate change is endangering the traditional life of communities and threatening the country's exceptional biodiversity.

Agricultural yields have diminished, the prevalence of malaria has increased, and marine ecosystems show signs of stress such as coral bleaching. Extreme weather increasingly provokes coastal and inland flooding as well as landslides. In the North Coast and Islands regions of Papua New Guinea, home to about 2.6 million people, coastal and inland flooding is the most dangerous climate-related hazard, threatening coastal communities, provincial towns and infrastructure. The country is home to the world's first climate refugees from the Carteret atolls, which are expected soon to disappear due to sea-level rise (Papua New Guinea, 2010).

The economy is dominated by the mining sector, which has seen substantial growth since 2000, but tourism is also increasing. Mining and oil and gas production generate significant CO_2 emissions, so for Papua New Guinea, mitigation is important. The government correctly recognizes that a rapid transition to renewable energy for electrical power generation should be the central component of mitigation policy. The focus for adaptation measures is concentrated on nine climate-related hazards (INDC, 2015):

1) coastal flooding and sea-level rise
2) inland flooding
3) food insecurity caused by crop failures due to drought and inland frosts
4) cities and climate change
5) climate-induced migration
6) damage to coral reefs
7) malaria and vector-borne diseases
8) water and sanitation
9) landslides.

The government received funding from the Adaptation Fund in 2012 for action focused on helping both the government and residents of the riverine North Coast and the coastal islands communities make informed decisions about how best to plan and respond to a coastal or inland flood. The project includes integrated riverbank protection measures across four provinces (Adaptation Fund, 2015). The UNDP has also funded similar action aimed at enhancing the adaptive capacity of communities in the same region.

What is missing is food security and fisheries. Whereas forest communities should generally be able to cope better with extreme weather, communities on the North Coast and the islands will be worse affected, and their dependence on fisheries means that measures to protect coastal and marine ecosystems are a priority.

Puerto Rico

The Commonwealth of Puerto Rico is an island territory of the US located in the Caribbean to the east of the Dominican Republic. It is part of an archipelago that includes the main island of Puerto Rico and a number of smaller islands including Mona, Culebra and Vieques. The main island's population is about 3.7 million; the capital is San Juan.

Tropical storms and hurricanes have become more intense over the last 20 years. Sea levels have risen by about 10 cm since 1960. Marshes have been submerged, mangroves inundated, and beaches eroded. Coastal flooding has worsened. At the same time, drought has become more common. In 2016 the drought was severe – several hundred thousand people faced water restrictions and in some areas the supply was shut off entirely for one or two days at a time (EPA, 2016).

Two-thirds of the island's electricity is generated from imported diesel fuel. The remaining third is generated by independent power producers using coal and natural gas (ETI, 2015). In 2015 there were 120 MW of wind power, 22 MW of solar PV power, and 21 hydropower units in operation, (ETI, 2015).

Being a US territory, Puerto Rico did not file an INDC submission and there is little information on adaptation planning. Almost all the reports document the multitude of threats and the observable impacts of a changing climate, but say very little about what is being done to mitigate these threats (PRCCC, 2013).

This is strange because Puerto Rico is hugely vulnerable to intensifying hurricanes fueled by the warming waters of the Caribbean. The Puerto Rico Electric Power Authority (PREPA)—which serves about half the population—filed for bankruptcy in July 2017. The island's electricity transmission and distribution systems are reportedly in terrible shape: dilapidated, poorly maintained, and fragile.

In September 2017, hurricane Maria smashed into the island virtually destroying the electricity transmission and distribution networks and flooding extensive areas of the capital. Only a few weeks before Maria, hurricane Irma also caused extensive damage. This scenario is going to be played out more frequently in the Caribbean as sea surface temperatures slowly rise. All the Caribbean islands, particularly Haiti, should pay close attention to what has happened in Puerto Rico.

Saint Kitts and Nevis, Saint Lucia, Saint Vincent and the Grenadines

These three small island countries, close neighbours in the eastern Caribbean, share the same climate and are threatened by the same impending dangers that the changing weather patterns will bring to their region. Part of the island chain of the Lesser Antilles, the countries are all volcanic islands with mountainous interior regions. St Lucia is the largest island (539 km^2) and the wealthiest, with a GDP almost twice that of the other two islands. It also attracts more tourists: over 300,000 a year compared with about 100,000 in St Kitts and Nevis and about 70,000 in St Vincent and the Grenadines. St Vincent, the poorest of the three countries, is distinct in that the country includes a chain of 31 smaller islands (the Grenadines) that stretch southwards from the main island.

Formerly dependent on agriculture as the mainstay of their economies, all the islands now depend on tourism to earn foreign exchange and provide employment. This characteristic means that the protection of coastal infrastructure and of coastal and marine resources should be a priority. Hotels provide government tax revenue and generate employment, but these businesses are often the worst offenders when it comes to coastal pollution. Strict regulation of hotels and other coastal service industries should be the first order of business for small island governments. Setbacks, solar water heaters, photovoltaic systems, and best-practice wastewater treatment facilities can all be mandated and regulated by island governments without the need for funding from donor agencies.

The islands of the eastern Caribbean are especially vulnerable to hurricanes. These violent storms may not become more frequent – expert opinion is divided – but they are likely to become more intense: powered by higher ocean temperatures and increased levels of moisture in the atmosphere. Storm surge coastal impacts will become more erosive and dangerous. At the same time, periods of drought have become more common, and the management of water much more of a problem. Few of the islands have vigorously promoted water harvesting by individual families, and yet this is a low-cost and effective measure. A reverse-osmosis plant on Bequia provides additional water to that island's small population, and this essential technology may need to be replicated on many of the other small islands in the Grenadines chain.

The INDCs submitted by the island governments provide few tangible details about how they will cope with more violent storms, unpredictable rainfall, periods of drought, failing agriculture, and fisheries under increasing stress. There are ways that small islands can strengthen their resilience to these threats, but investing huge amounts of time in long-winded consultative meetings that produce yet more road maps, national strategies, policy documents, and wish-list action plans is not the way to go.

Samoa

Samoa is a small island country located in the southwest Pacific, consisting of four main inhabited islands and six smaller, uninhabited islands, all of volcanic origin. The islands are therefore characterized by a rugged and mountainous topography.

On the smaller of the two main islands, Upolu, the central mountain range runs along the length of the island with peaks rising more than 1000 metres above sea-level. Savai'i, the larger island, has a central core of volcanic peaks reaching 1858 metres at the highest point, Mt Silisili. The population is about 190,000 and the total land area is 2840 km^2. The capital, Apia, is in the northern part of Upolu, the smaller island.

Like all the Pacific SIDS, Samoa is vulnerable and exposed to climate change threats. Sea-level rise will exacerbate coastal erosion, loss of land, and the dislocation of island settlements. Coastal floods are also likely to become more frequent and severe. Cyclones are a constant threat: when Cyclone Evan hit Samoa in 2012, 7500 people were affected, and about 2000 houses were damaged. The cyclone also badly damaged two of the five hydropower stations operating at the time. Cyclones may not become more frequent, but all the data suggests that they will become more intense and therefore more destructive.

Samoa has embarked on a vigorous programme accelerating the transition to renewable energy, with the target being 100% renewable energy in 2017. This policy is coupled with a strong push to increase energy efficiency, so this commendable policy will definitely reduce the islands' GHG emissions. This initiative is well-timed and well-focused.

But when it comes to the need for adaptation measures aimed at reducing the country's vulnerability and increasing the resilience of the island's population to climate change impacts, there is a lack of focus and seemingly very little action (INDC, 2015). The country needs a clear strategy and action plan to implement adaptation measures focused on the protection of coastal zones and marine biodiversity, integrated water resources management, and food security.

São Tomé and Príncipe

The volcanic islands of São Tomé and Príncipe are located in the Gulf of Guinea off the west coast of Africa. It is one of Africa's smallest countries.

São Tomé is the more mountainous of the two islands. Its highest point, Pico de São Tomé, rises to over 2000 m, and the Pico Cão Grande (Great Dog Peak) is a landmark feature of the island. The island of Príncipe, although smaller is almost as mountainous.

The country is poor: two-thirds of the population of about 200,000 live below the poverty line of US$3 a day (World Bank, 2016). Only half the population have access to electricity, and only a third have proper sanitation facilities. The main economic activities involve tourism, transport, communication and construction. Agriculture and fishing are economic mainstays for the majority of the population, despite the modest contribution of these activities to GDP.

The lack of electricity means that rural families rely on kerosene for lighting and wood for cooking. Even in urban areas, wood is used for cooking. The smoke generated from wood fires creates serious health issues for women and children, as well as huge pressure on forest resources: only about a quarter of the islands are forested.

Lying close to the Equator, the islands are not at high risk from hurricanes. The main threats are due to the lower levels of rainfall and the increasing incidence of drought. The dry season, 'gravana1', has been growing longer. At the same time, there are increasing incidents of torrential rain. The management of water resources is increasingly a problem, including for the hydropower installations that need adequate river flow. One of the two hydropower installations, Guégué, has been out of operation since 2011 (World Bank, 2016).

The lack of electricity on the islands is a serious problem that should be an urgent priority for the international finance institutions and aid agencies. Communities without electricity are enormously vulnerable to extreme weather events, particularly at night.

Electricity is also essential for the development of the tourism industry, one of the few economic lifelines available to the country. This focus suggests that the protection of coastal and marine resources should be an absolute priority, not just for the coastal communities that rely on fishing, but also for the developing tourism sector which flourishes when there are pristine beaches, healthy coral reefs, and well managed marine ecosystems (NAP, 2006).

Seychelles

The Republic of Seychelles is a multi-island country in the Indian Ocean about 1000 km north of Madagascar. The area of the island is 457 km^2, and with a population of about 92,000 people the country is the least populated of any African state.

The Seychelles has developed from a largely agricultural society to a market-based diversified economy, with agriculture being overtaken by a rapidly rising service sector as well as tourism. Since 1976, GDP per capita has increased nearly seven-fold. Despite the country's new-found economic prosperity, poverty remains widespread due to a high level of income inequality.

The main climate change threats facing the Seychelles are similar to those facing the majority of small island states: changes in rainfall patterns leading to flooding and landslides, extended periods of drought, increasing sea temperatures, changes in acidity and damage to marine ecosystems, more intense storms and storm surges, and sea-level rise over the longer term. However, what distinguishes climate change policy in the Republic of Seychelles is a clear statement that mitigation is a secondary objective for the country. As the INDC states:

> *Given that the Republic of Seychelles is a net sink, its contributions to climate change mitigation to contribute towards the objectives of the UNFCCC will be the co-benefit of enhancing its energy security and reducing its energy bill.*
>
> (INDC, 2015)

The report goes on to say:

> *In Seychelles, climate change mitigation to stabilize the climate system is not a primary objective. Mitigation is rather seen as an important outcome or by-product of decreasing the country's dependence on imported fossil fuels (i.e. an increase in energy security), and to enhance its balance of trade profile (through a reduction in its energy bill).*

This is a smart policy that the majority of the SIDS should heed. Even those islands that are not net sinks of CO_2 should focus their efforts on transitioning to renewable energy systems, because setting and achieving renewable energy targets not only reduces the foreign exchange burden of imported petroleum fuels and increases energy sector resilience, but it substantially cuts back emissions of CO_2. It is a win-win situation that more of the SIDS should have identified and articulated in their INDC communications.

The adaptation strategies outlined by the Seychelles to address the primary vulnerabilities are as follows:

- *Critical infrastructure*: Climate change adaptation to be mainstreamed in all sectors with critical infrastructure. Planning process for all new developments with associated improvements in the building codes and their rigorous enforcement.
- *Tourism*: Greater co-management of the sector by the Ministry of Tourism and Department of Risk and Disaster Management, as well as with the Ministry of Environment, Energy and Climate Change.
- *Food security*: A sustainable modern agriculture supported by new and innovative technologies across all food production supply and value chains, and by skilled and qualified human resources, and integrated with the *Blue Economy* and *Seychelles Strategic Plan 2015*.
- *Biodiversity*: *Seychelles Biodiversity Strategy and Action Plan*. Fully implemented and enforced Biodiversity Law, fully bio-secure border.
- *Water security*: Fully integrated approach to water security that addresses issues such as ecosystem health, waste management, water treatment and supply, sewage, agriculture, and so on.
- *Energy security*: More resilient energy base with greater innovation of renewable energy where practicable. Efficient fuel-based land transport and more use of electric vehicles charged with renewable energy technology. Strengthened cooperation between Government entities.
- *Health*: Health sector able to respond to population increase and its additional climate-related health burden. Exploration of relevant potential science and technology innovations.
- *Waste*: Waste managed according to strict hierarchy and waste policy fully implemented. Exploration of relevant potential science and technology innovations.

The importance of establishing energy efficiency targets as a means of further reducing greenhouse gas emissions has also been recognized. The National Energy Efficiency programme aims to:

- Promote energy-efficient appliances to achieve 10% energy savings by 2035.
- Promote solar hot water heaters with a target of 80% use in households and 80% in services by 2035.
- Apply new regulations on the use of air-conditioning equipment, with a target of 20% energy savings in the service sector.
- Introduce a new building code for household dwellings (featuring natural ventilation, roof insulation, etc.) with a target of 50% energy savings on fans and AC in households by 2035.
- Promote co-generation (production of hot water and water heat from electricity generation) in hotels, with a target of covering 20% of hot water needs by 2035.

The INDC (2015) prepared by the Republic of Seychelles is one of the best communications to the UNFCCC prepared by any of the SIDS.

Singapore

The Republic of Singapore is the world's only island city-state. Situated just north of the Equator, the low-lying territory consists of the main island together with over 60 islets. Since independence in 1959, extensive land reclamation has increased its area by 23%, but the state remains one of the most densely populated in the world: in 2014 the population density was estimated as 8138 persons per km^2 (UNDESA, 2016).

A sophisticated and technically advanced island state, Singapore has monitored and analyzed climate change impacts in considerable technical detail. The first study was completed in 2013; the second (in collaboration with the UK Met Office Hadley Centre) was completed in 2015. What is known is that from 1972 to 2014, the annual mean temperature increased from 26.6 to 27.7 °C. Sea levels are rising at about 1.7 mm/year. Rainfall patterns are changing: the maximum rainfall intensity recorded each year has increased from 80 mm in 1908 to 107 mm in 2012 (INDC, 2015).

Sea-level rise is the most immediate threat to the island. Much of the country lies only 15 m above mean sea-level, with about 30% of the island lying less than 5 m above mean sea-level. But water resources, biodiversity, public health, heat island effects, and food insecurity are all major concerns.

Singapore's key adaptation measures already well underway include:

- *Food security*: Diversifying food sources, local production of key food items and rice stockpiling.
- *Infrastructure resilience*: Strict oversight of infrastructure through the Building Control Act. Flood barriers have been installed at subway stations. Energy and telecommunications installations must meet performance standards.
- *Public health*: The State has an integrated regime of environmental management to suppress the mosquito population. Early warnings of increased risk are in place.
- *Flood management*: Investments in drainage infrastructure have reduced flood-prone areas 100-fold. All commercial, industrial, institutional and residential developments

greater than 0.2 ha are required to implement on-site water detention measures to reduce the peak discharge into the public drainage systems.

- *Water security*: Strengthening the resilience of the sector focuses on local catchment, imported water, NEWater (recycled), and desalination. Managing the supply and demand for water is a major element of an adaptation programme.
- *Coastline protection*: More than three-quarters of Singapore's coastline is protected by hard structures such as sea walls and stone embankments. The remaining coastline is sandy beaches and mangrove swamps. Since 1991, all new coastal lands have been reclaimed to 1.25 m above the highest recorded tide level. In 2012 this security margin was increased to 2.25 m above the highest tide mark.
- *Safeguarding biodiversity*: The island has implemented the vision of a 'City in a Garden' that encompasses an extensive network of parks and natural green spaces.
- *Regional climate modelling*: The country has established the Centre for Climate Research Singapore (CCRS) in order to strengthen its scientific capacity to predict climate change tendencies and weather extremes.

The one area of adaptation where Singapore falls short is the transition to renewable energy. Almost totally dependent on imported liquid fuels for power generation, the government cites population density, limited land space, and a poor wind regime as reasons for its lack of progress towards reducing its dependence on fossil fuels. But wind power can be offshore, and photovoltaic power can be installed on rooftops and even floated on reservoirs. Much more could be done with a little more imagination and forcefulness.

Solomon Islands

The Solomon Islands is a multi-island country consisting of six main islands and over 900 smaller islands in the south Pacific Ocean east of Papua New Guinea. The total land area is over 28,000 km². The country's capital, Honiara, is located on the island of Guadalcanal. The country takes its name from the Solomon Islands archipelago, which is a large group of Melanesian islands that includes the North Solomon Islands.

The six main islands are of volcanic origin and are rugged and mountainous. Between and beyond the larger islands are hundreds of small volcanic islands and low-lying coral atolls. All the mountainous islands are forested, with many coastal areas surrounded by fringing reefs and lagoons.

The islands are situated within the earthquake belt or 'ring of fire', which makes the country vulnerable to earthquakes and tsunamis. Cyclones are also a threat. Tropical storms are frequent, with rainfall levels reaching more than 250 mm in a 12-hour period. One such event in 2014 caused widespread flooding and damage along the Mataniko River (INDC, 2015).

Around 80% of the population of almost 600,000 people live on low-lying areas, so the country is extremely vulnerable to sea-level rise, storm surges and ocean swells. Several vegetated reef islands have vanished during the last 60 years, while others have experienced severe coastal shoreline recession which has destroyed coastal villages (Albert *et al.*, 2016).

Many families practice subsistence agriculture supported by fishing and forest products. Tourism is modest: only 20,000 visitors were recorded in 2014 (UNDESA, 2015).

It is almost impossible to protect over 900 islands against sea-level rise, extreme weather and climate change. The country needs to develop a realistic strategy that focuses on the protection of the larger islands where the majority of the population will eventually have to reside. A swift transition to renewable energy and a priority programme of rural electrification on the main islands will build resilience and provide incentives for islanders from outlying islands to relocate.

Given the levels of precipitation, the management of water does not seem to be an urgent priority – better to focus on ecosystem-based adaptation programmes on the coastal zones of the larger islands. Above all, the fisheries must be protected; this means reducing stress on coral reefs from wastewater pollution and other sources of coastal pollution.

The Solomon Islands' INDC communication notes the importance of the Green Climate Fund's 'readiness' facility. The country should move rapidly to access this support.

Suriname

Suriname is a small country located between Guyana and French Guyana on the northeast coast of the South American continent. The country is heavily forested and most of the population is located along the northern coast and in the capital Paramaraibo. The coastal zone is low-lying, marshy, and vulnerable to extreme weather and storm surge flooding. Being close to the Equator, hurricanes are infrequent but tropical storms are common.

Given the vulnerability of the coast, the government has considered moving the capital Paramaraibo further inland, although this is recognized as an expensive and complex project (INDC, 2015).

Climate change adaptation measures are understandably focused on the coastal zone. Hard measures include dykes and drainage systems for urban and non-urban areas, and ecosystem-based measures including the protection, restoration and expansion of mangrove forests. Improvements in water resources management, protection of freshwater resources in aquifers and rivers, promotion of sustainable land management, applying innovative technologies in the use of land, and instituting measures towards increasing ecosystem resilience to ensure that they adapt to the changing climate, are all actions being implemented (INDC, 2015).

However, given the low elevation of the coastal zones and the huge areas of land available in the interior of the country, it may make sense over the longer term for the government gradually to move government offices and service infrastructure to towns further up-river from the coast.

Ecotourism is a significant industry and may continue to develop. This is not beach tourism, and so infrastructure should be built inland and on higher ground. Suriname has the luxury of space: there is plenty of room to move infrastructure and communities gradually to higher elevation and better-protected areas.

Suriname must protect its forests. Not only are they an important global carbon sink: they are a treasure trove of biodiversity and the key to the development of a world-class ecotourism industry that should substantially benefit the indigenous population, if properly managed.

Timor-Leste

Timor-Leste (or more commonly 'East Timor') is a country occupying the eastern half of the island of Timor and the nearby islands of Atauro and Jaco. The country also includes a small region of West Timor called Oecusse.

The country only achieved independence in 2002, after a bloody war with Indonesia that left the country half-destroyed and impoverished.

The main climate threats to Timor-Leste are floods, landslides, drought, and tropical cyclones. Half the population is engaged in agriculture, but the sector only contributes 4% to GDP – a sure sign that agriculture is subsistence and only marginally productive. As a result, food security is a perennial problem with almost half of the population counted as food insecure or vulnerable to food insecurity. This insecurity is caused by political unrest, poverty, poor soils, and inefficient farming practices, but increased climate variability and extreme weather exacerbate the situation. Rising sea-level threatens the capital Dili with flooding, especially during storm surges.

Although the country is heavily forested, slash-and-burn agriculture has reduced forest cover from 65% in 1990 to 48% in 2014.

Timor-Leste did not file an INDC communication, and so little is known about the government's plans for mitigation and adaptation measures. But like so many of the SIDS, expending time and effort on mitigation efforts to reduce CO_2 emissions is a wasteful distraction for a country whose emissions are negligible. In contrast, planning for climate change impacts and putting in place measures to protected poor farming and coastal communities is absolutely a priority.

The coastal population of Timor-Leste is particularly vulnerable to the threats of sea-level rise and extreme weather. Approximately 40% of the population lives in the coastal areas. They are not only susceptible to flash floods and landslides originating from the upstream hill areas, but are also likely to face increased incidents of storm surge and coastal flooding. These hazards have serious impacts on coastal and marine ecosystems, especially for mangroves, estuaries and coral reefs, which are already under stress because of coastal zone development and population growth. The coastal zone (and habitats) of Timor-Leste, for example, are already subject to a high degree of exploitation and resource use. Increased storminess, wave swells, sea surges and sea-level rise will threaten the country's still fragile development gains.

Coastal impacts are likely to include (UNDP, 2015):

- increased groundwater contamination by salt water intrusion;
- direct damage to physical infrastructure and disruptions in water supply services, road networks, buildings and port operations as a result of coastal submersion;
- damage and disruption to coastal power stations and transmission infrastructure through erosion, flooding and saltwater damage;
- flooding and destruction of coastal settlements;
- direct physical damage to forest and terrestrial ecosystems;
- reduced health, diversity and productivity of coastal and inshore marine ecosystems and species;
- loss or destruction of coastal vegetation, species and habitats;

- physical damage to coral reefs and mangroves by strong wave action;
- increased erosion of beaches, shorelines and coastal land, loss of breeding and nesting habitats;
- increased damage to and destruction of bridges, roads, roadsides, culverts, drainage structures and river embankments;
- increased risk of damage to offshore oil and gas infrastructure and disruption to operations;
- increased risk of accidents, spills, leaks and pollution resulting from flooding and wind damage to fuel storage facilities and other installations housing hazardous materials;
- damage to schools, homes and community buildings such as churches and health clinics.

Timor-Leste requires immediate and substantial help from the UNDP, the Green Climate Fund Readiness Facility, and the bilateral donors, to help the country to get back on its feet. The country needs a long-term programme of technical assistance.

Tonga

The island kingdom of Tonga, located in the central southern Pacific, consists of four clusters of islands stretching north–south: Tongatapu and 'Eua in the south, Ha'apai in the centre, Vava'u in the north, and Niuafo'ou and Niuatoputapu in the far north. The capital Nuku'alofa is on Tongatapu, the largest island.

A total of 169 islands, of which 36 are inhabited, make up the kingdom. The total surface area is about $750\,km^2$ dispersed over some $700,000\,km^2$ of ocean. Tonga has a population of just over 100,000 people, of which about 70% reside on Tongatapu.

The archipelago is situated close to the so-called 'ring of fire', a zone of frequent seismic activity and tsunami. Most of the atoll islands, including the main island, are very flat with an average altitude of 2–5 metres, and so Tonga is highly vulnerable to sea-level rise, storm surges, and tsunami inundation (INDC, 2015).

The INDC document provides little information on the measures to be implemented by the government to combat climate change impacts and to build resilience. The planning for implementation is to be set out in a new Joint National Action Plan for Climate Change Adaptation and Disaster Risk Reduction (JNAP), together with other plans (at sector, island, and community levels) that are to be fully aligned with the goal and targets of the policy (INDC, 2015).

Climate change policy for Tonga has been outlined in a 2016 report which aims at a 'more resilient Tonga by 2035' (TCCP, 2016). However, the report is all about the process and the principles, not about action. While process, principles and policy are important, it is late in the day still to be engaged in exhaustive consultative meetings in order to produce a policy document that all stakeholders can agree on.

The policy document sets out 20 targets for a 'resilient Tonga'. These targets now need to be translated into the quantitative results expected from concrete and specific actions that are formulated in a way that will facilitate funding from the major climate funds: the Green Climate Fund and the Adaptation Fund.

Trinidad and Tobago

Trinidad and Tobago is the most industrialized country in the English-speaking Caribbean. It is the leading Caribbean producer of oil and gas, and its economy is mainly based on the exploitation of these hydrocarbon resources. Oil and gas is the leading economic sector, accounting for 40% of GDP and 80% of exports. Agriculture, once an important economic sector, now contributes less than 0.5% to GVA and the sector employs less than 4% of the workforce (INDC, 2015).

Nevertheless, the impacts of a changing climate are discernible: the country has experienced increases in mean temperatures over the last three decades, and like all the SIDS, the country is vulnerable to sea-level rise, variable and unpredictable precipitation, increased intensity of storms, and hillside and coastal erosion. Although the islands of Trinidad and Tobago are the most southern of the Caribbean islands, and thus generally spared the impact of the Caribbean hurricanes passing to the north, the intensity and frequency of tropical storms is expected to increase.

The country generates all its electricity from natural gas and aims to achieve greater efficiency by installing combined cycle generation at all power plants. Perhaps understandably, the country's INDC is all about reducing CO_2 emissions; there is almost nothing about adaptation (NCCP, 2011; INDC, 2015).

However, the island of Tobago, further to the north, is much more vulnerable to climate change impacts than the larger more industrialized island of Trinidad. It has substantial tourism infrastructure on its many fine beaches, and this infrastructure is vulnerable to storm surges, sea-level rise and coastal erosion. Adaptation measures should focus on protecting Tobago's marine resources and coral reefs, employing ecosystem-based approaches. The whole of the island of Tobago, including all the coastal and marine inshore areas, should be mapped, zoned and protected.

Tuvalu

Tuvalu, formerly known as the Ellice Islands, is a Polynesian island nation located in the Pacific Ocean, about midway between Hawaii and Australia. The archipelago consists of nine small islands, six of them being atoll islands (with ponding lagoons) namely Manumea, Nui, Vaitupu, Nukefetau, Funafuti and Nukulaelae. The remaining three islands (Nanumaga, Niutao and Niuliakita) are raised limestone reef islands. All the islands are less than five metres above sea-level, with the largest island, Vaitupu, having a land area of just over $5\,km^2$ – hardly large enough to build an airport. The total land area, home to about 10,000 people, is about $26\,km^2$.

Tuvalu is a LDC with a per capita income of less than $4000. It is the smallest of the SIDS. Due to the lack of land area and resources, the scope for economic diversification is minimal. Almost everything is imported: fuel, food, skilled labour and services. There is virtually no tourism – only 1000 visitors arrived in 2014 (UNDESA, 2016).

Tuvalu's future climate will cause increasing hardship for the islanders. The forecast made by the Pacific Climate Change Science Program could apply to almost all the Pacific SIDS (PCCSP, 2011):

- Temperature will continue to increase
- More very hot days will occur
- Rainfall patterns will change and become less predictable

- More extreme rainfall days
- Less frequent but more intense tropical cyclones
- Sea-level will continue to rise
- Ocean acidification will increase.

Tuvalu's CO_2 emissions are less than 0.000005% of global emissions – apparently too small for the UNDESA database even to register them. Why then does Tuvalu's INDC only talk about the need to reduce them further? There is nothing about adaptation in the INDC (2015), even though the need to cope with the strengthening climate change impacts on the islands is imperative.

Poor island communities depend on the rain and the sea to survive. Although the outlook is bleak, Tuvalu should get organized to collect, store and manage every drop of its rainfall. It must build community centres that protect families from violent storms. It must find ways to nurture and protect its marine biodiversity.

The policy to transition to 100% renewable energy for power generation is an excellent one. But the government should also ensure that all families have access to electricity and to communication systems. Connected is protected.

US Virgin Islands

The US Virgin Islands are located in the Caribbean to the east of Puerto Rico. The three main islands are: St Thomas, St Croix, and St John. The smaller Water Island is also part of the group. The islands are an unincorporated territory of the US with policy relations between the Virgin Islands and the US under the jurisdiction of the US Department of the Interior. The country therefore did not file an INDC communication.

Tourism, trade and other services are the primary economic activities, accounting for nearly 60% of the country's GDP and about half of total civilian employment. The islands host nearly 3 million tourists per year, mostly from visiting cruise ships. The agriculture sector is small, with most food being imported. Industry and government each account for about a fifth of GDP. The manufacturing sector consists of rum distilling, electronics, pharmaceuticals and watch assembly. A refinery on St Croix, one of the world's largest, processed 350,000 barrels of crude oil a day until it was shut down in February 2012, after operating for 45 years.

Like all the Caribbean islands in the Antilles chain, the Virgin Islands are vulnerable to extreme weather and hurricanes. Before Maria in September 2017, the most powerful storm was hurricane Hugo, which devastated the islands in 1989, causing catastrophic physical and economic damage. Hurricanes and tropical storms continue to be a severe threat throughout this region of the Caribbean.

Adaptation measures aimed at building resilience to the impacts of climate change should focus first on vulnerable coastal communities. A workshop organized by the Nature Conservancy in 2014 identified ten coastal areas in the territory most vulnerable to climate change. They were: Two Brothers, Demarara, Kings Quarter, Honduras, Nadir, East Street, Mount Pleasant, Retreat, Bovoni and Enighed (Schill *et al.*, 2014). The workshop proposed measures aimed at:

- improving coastal protection;
- increasing emergency services to particular areas; and
- building community's resilience to hazardous weather and future impacts.

A chronic problem on Caribbean islands is ensuring adequate supplies of fresh water. Desalination plants are becoming the norm. So too is the transition to renewable energy power generation. Both of these technologies are being promoted and extended in the US Virgin islands.

Vanuatu

The Republic of Vanuatu is an island nation located in the western Pacific. The country is an archipelago of over 80 islands stretching 1300 km from north to south. The largest towns are the capital Port Vila, on Efate, and Luganville on Espiritu Santo. The highest point in Vanuatu is Mount Tabwemasana, 1879 m high on the island of Espiritu Santo.

The islands are mostly mountainous with narrow coastal plains. The larger islands are characterized by rugged volcanic peaks and tropical rainforests. The archipelago is located in a seismically and volcanically active zone, and the islands are exposed to hazards including volcanic eruptions, earthquakes, tsunamis and landslides. However, because of their volcanic origin, the islands of Vanuatu are much better protected from sea-level rise and extreme weather than many low-lying island states in the Pacific.

The population is over a quarter of a million. Tourism is an important economic sector: over 100,000 tourists visit the islands each year. But agriculture is still important, accounting for 26% of GVA in 2014 and the majority of employment (UNDESA, 2016).

Climate change and changing weather patterns are already having an impact. Agricultural production, fisheries, human health and tourism are all expected to suffer (INDC, 2015).

In its National Adaptation Program of Action, the government of Vanuatu identified five priority areas of action:

1) Agriculture and food security
2) Sustainable tourism development
3) Community-based marine resource management
4) Sustainable forest management
5) Integrated water resources management.

These are the key priorities for the majority of the small island states. The Least Developed Country Fund is providing financial support for agriculture, food security and water resources management (INDC, 2015).

Multi-island country governments are faced with the particular problem of how to manage so many islands spread over large areas of ocean. Each inhabited island should have its own climate change adaptation plan. This requires a major investment of time and technical resources, and the international funding agencies (the Green Climate fund and the Adaptation Fund) should prioritize financial and technical assistance to the multi-island SIDS.

References

Adaptation Fund (2014) *Adaptation Story: Cook Islands*. Adaptation Fund, Bonn.
Adaptation Fund (2015) *Analysis of Climate Change Adaptation Reasoning in Project and Programme Proposals Approved by the Board*. AFB/PPRC.17/5. Available at www.

adaptation-fund.org/wp-content/uploads/2015/12/AFB.PPRC_.17.5-Analysis-of-climate-change-adaptation-reasoning_final_Nov...pdf

Albert, S., Leon, J.X., Grinham, A.R., *et al.* (2016) Interactions between sea-level rise and wave exposure on reef island dynamics in the Solomon Islands. *Environmental Research Letters*, 11, 054011.

Al-Zubari, W. (2014) *The Costs of Municipal Water Supply in Bahrain.* Chatham House, The Royal Institute of International Affairs, London.

BREA (2015) *Consumer Guide Solar PV.* Barbados Renewable Energy Association. Available at www.barbadosenergy.org

British Virgin Islands (2011) *The Virgin Islands' Climate Change Policy: Achieving Low-carbon, Climate Resilient Development.* Government of the Virgin Islands, Road Town.

CCPF (2015) *Climate Change Policy Framework for Jamaica.* Government of Jamaica, Kingston.

EPA (2016) *What Climate Change Means for Puerto Rico.* Environment Protection Agency, Washington, DC.

ETI (2015) *Energy Transition Initiative: Island Energy Snapshots.* Department of Energy, Office of Energy Efficiency and Renewable Energy, Washington, DC. Available at https://www.energy.gov/eere/island-energy-snapshots

Fletcher, C.H. and Richmond, B.M. (2010) *Climate Change in the Federated States of Micronesia: Food and Water Security, Climate Risk Management and Adaptive Strategies.* Center for Island Climate Adaptation and Policy, University of Hawai'i, Honolulu.

GEAP (2013) *Guam Energy Action Plan.* National Renewable Energy Laboratory, Denver, CO.

GFDRR (2012) *Disaster Risk Management in Latin America and the Caribbean Region: GFDRR Country Notes: Belize.* Global Facility for Disaster Reduction and Recovery, World Bank, Washington, DC.

HDR (2015) *Human Development Report.* United Nations Development Programme, New York.

INDC (2015) The INDCs for all the Small Island Developing States that made submissions can be found on the UNFCCC website at www4.unfccc.int/submissions/INDC/Submission%20Pages/Submissions.aspx

IRENA (2013) *Renewable energy opportunities and challenges in the Pacific Islands region: Cook Islands.* Available at www.irena.org/DocumentDownloads/Publications/Cook-Islands.pdf

JNAP (2012) Niue's Joint National Action Plan for Disaster Risk Management and Climate Change. Department of Environment, Government of Niue, Alofi, Niue.

KJIP (2014) Kiribati Joint Implementation Plan on climate change and disaster risk management. Available at www.pacificdisaster.net/pdnadmin/data/original/KIR_2014_Joint_implementation_plan_for_CC_and_DRM.pdf

NAP (2006) *National Adaptation Programmes of Action on Climate Change.* Republica democratica de S. Tome E. Principe, Ministerio des recursos naturalis e ambiente, Sao Tome.

NCCP (2011) *Government of the Republic of Trinidad and Tobago, National Climate Change Policy.* Available at www.ema.co.tt/new/images/policies/climate_change_2011.pdf

Ochs, A., *et al.* (2015) *Caribbean Sustainable Energy Roadmap and Strategy: C-SERMS.* Baseline report and assessment. Worldwatch Institute, Washington, DC.

Papua New Guinea (2010) *Papua New Guinea's commitment to act on climate change.* Papua New Guinea Office of Climate Change and Development, Port Moresby.

PCCP (2015) Pacific Climate Change Portal. Available at http://www.pacificclimatechange. net/country/american-samoa

PCCSP (2011) *Pacific Climate Change Science Program. Current and future climate of Tuvalu.* Available at www.pacificclimatechangescience.org/wp-content/ uploads/2013/06/4_PACCSAP-Tuvalu-10pp_WEB.pdf

PRCCC (2013) *Puerto Rico's State of the Climate: Assessing Puerto Rico's social-ecological vulnerabilities in a changing climate. Executive Summary 2010–2013.* Department of Natural and Environmental Resources, San Juan.

RISA (2016) Pacific RISA: Northern Mariana Islands. Available at http://www.pacificrisa. org/places/commonwealth_of_the_northern_mariana_islands/

Schill, S., Brown, J., Justiniano, A. and Hoffman, A.M. (2014) *US Virgin Islands climate change ecosystem-based adaptation: Promoting resilient coastal and marine communities.* The Nature Conservancy, Arlington, VA.

Simpson, M.C., Clarke, J.F., Scott, D.J., *et al.* (2012) *CARIBsave Climate Change Risk Atlas (CCCRA) – Anguilla. Summary document.* DFID, AusAID and the CARIBSAVE partnership, Barbados, West Indies.

SOPAC (2011) *Desalination in Pacific Island Countries: A preliminary overview.* SOPAC Technical Report 437, Suva, Fiji.

SPC (2015) *Palau Climate Change Policy: For Climate and Disaster Resilient Low Emissions Development.* Government of Palau, Secretariat of the Pacific Community, Suva, Fiji.

SPREP (2017) Secretariat of the Pacific Regional Environment Programme – climate change overview. Available at http://www.sprep.org/Climate-Change/ climate-change-overview

SREP (2015) *Scaling up renewable energy program: SREP Investment plan for Haiti.* Climate Investment Funds. Available at www.climateinvestmentfunds.org/sites/default/ files/Haiti%20SREP%20TOR%20Joint%20Mission.pdf

TCCP (2016) *Tonga Climate Change Policy: A Resilient Tonga by 2035.* Department of Climate Change, Ministry of Meteorology, Energy, Information, Disaster Management, Environment, Climate Change and Communications, Nufu'alofa.

UNDESA (2015) *World Pocketbook 2015.* UN Department of Social and Economic Affairs, New York.

UNDESA (2016) *World Pocketbook 2016.* UN Department of Social and Economic Affairs, New York.

UNDP (2015) *Terms of reference: Building shoreline resilience of Timor-Leste to protect local communities and their livelihoods.* Available at www.thegef.org/sites/default/files/ project_documents/Project_document_rev._0.pdf

UNESCO (2006) *The Use of Desalination Plants in the Caribbean.* Documentos Tecnicos del PHI-LAC No. 5. Caribbean Environmental Health Institute, Paris.

USGS (2016) *Many atolls may be uninhabitable within decades due to climate change.* Available at www.usgs.gov/news

Wade, S., Leonard-Williams, A. and Salmon, K. (2015) *Assessing climate change and its likely impact on selected UK Overseas Territories: Inception Report.* UK Meteorological Office, London.

WF (2015) *World Factbook 2015.* US Government, Central Intelligence Agency, Washington DC.

World Bank (2016) *Project Appraisal Document for Sao Tome and Principe for a power sector recovery project.* Report No. PAD1736. World Bank, Washington DC.

Index

Climate Change Adaptation in Small Island Developing States, First Edition. Martin J. Bush.
© 2018 John Wiley & Sons Ltd. Published 2018 by John Wiley & Sons Ltd.